Transdisciplinary Systems Engineering

Azad M. Madni

Transdisciplinary Systems Engineering

Exploiting Convergence in a Hyper-Connected World

Foreword by Norman R. Augustine

Springer

Azad M. Madni
Viterbi School of Engineering
University of Southern California
Los Angeles, California, USA

ISBN 978-3-319-62183-8 ISBN 978-3-319-62184-5 (eBook)
DOI 10.1007/978-3-319-62184-5

Library of Congress Control Number: 2017947157

Printed on acid-free paper

This Springer imprint is published by Springer Nature
The registered company is Springer International Publishing AG
The registered company address is: Gewerbestrasse 11, 6330 Cham, Switzerland

This book is dedicated to my parents, Taher and Sara, and my mentor, Professor John Lyman of UCLA. They were and still are a source of inspiration to me.

Foreword

When a butterfly flaps its wings in Mexico, can it cause a hurricane in China? When a tree falls in Ohio, can it shut off electric power throughout the northeastern United States and part of Canada?

Scientists and philosophers still debate the former, but we know the answer to the latter with certainty. It of course actually happened in 2004, leaving some 50 million people in the dark.

These are examples of complex systems engineering problems – the topic of this well-written and thoughtful book by my colleague, Azad Madni. For the expert, this book is filled with insightful assessments of real-world issues and explanations of complex systems engineering techniques. For the more general reader, it offers systems engineering principles for applications ranging from autonomous vehicles to baseball.

Dr. Madni deftly tackles the most intransigent forms of complex systems: nonlinear, transdisciplinary systems. Such systems, at their best, can accomplish miraculous things, for example, transporting by air the equivalent to the population of Houston throughout the United States each day so safely that a passenger can travel the equivalent of circumnavigating the Earth 40,000 times without, on average, becoming a fatality. But, at their worst, they can lead to Fukushima-like tragedies. This book steers the designer, builder, operator, and user of complex systems away from the latter and toward the former through transdisciplinary approaches that exploit convergence among disciplines.

The challenge, of course, is that most real-world systems are, in fact, complex. Systems with less than a dozen elements, each of which can interact with each of the others, can have more states than there are stars in our galaxy. One of many consequences of this is that truly complex systems cannot be fully tested. The implications of this were demonstrated early in the US space program when a spacecraft designed for a planetary flyby of Venus made it only a few miles from the launch pad. The culprit: one erroneous symbol among hundreds of thousands of symbols in the software code in a backup, untested sequence that, of course, was entered during the spacecraft's actual flight. Systems engineers soon learn that Mother Nature is not malevolent – but is very unforgiving. Thus, the importance of this book. It offers several examples of cross-disciplinary approaches to solving problems that appear intractable when viewed through the lens of a single discipline.

Additionally, complex systems are increasingly interdisciplinary, posing further challenges to those who would design them. The nation's energy grid, for example, involves electrical systems, chemical systems, mechanical systems, structural systems, and more. Worse yet, they involve, yes, *humans!* This book offers several key insights on factors that need to be taken into account when attempting to maximize joint human-system performance. But more on that in a moment.

Perhaps a good place to begin is to ask, "What is a system?" The author answers this question elegantly. A minimalist definition of a system is a collection of two or more elements that interact with one another. But matters quickly become more complicated once we begin to address real-world problems. To a jet engine designer, the "system" is the engine, with its fans, controls, fuel, cowling, etc. However, to an aeronautical engineer, the system includes the entire aircraft: engines, wings, tail, fuselage, and, yes, even the pilot. But to a transportation engineer, the "system" also includes air terminals, air traffic control, runways, maintenance facilities, and passengers.

This is where a major challenge arises: when designing or analyzing a system, what should be included in "the system," and what can safely be excluded? To include "everything" makes the system impervious to analysis. However, to exclude a key element can make the system vulnerable and potentially hazardous. In the above example, is weather within the system's boundary, or is it more appropriately considered to be part of the environment? And what about the process of weather *prediction*? Or how should one treat the possibility of a meteorite damaging a spacecraft or a solar flare affecting communication links?

For totally automated or autonomous vehicles, states in the United States place major design constraints in the form of regulations; for example, some states require that all highway vehicles have a steering wheel and a brake pedal. So, how should such regulations be treated when defining a system?

Einstein offers some very succinct overarching advice: "Make everything as simple as possible, but not simpler!"

Early systems principally comprised physical elements. Then, something called software appeared. In my book, *Augustine's Laws,* I note that software is a colorless, odorless, weightless substance that, like entropy, always increases! The same could be said of today's complex systems which can have millions of lines of code, any one of which, when improperly conceived or written, can impede or compromise the proper functioning of a system. Witness the occasional shutdown of airline operations due to a "computer problem" or the unintended consequences thereof on vehicle travel near the airport? The culprit is often what in electronic systems is referred to as a "sneak circuit," a failure pathway that was not recognized – but should have been!

But here, the big problem in complex systems enters: *humans*! Yes, humans are an integral part of most systems and are the most unpredictable element in those systems. Believe it or not, humans are not always rational. Numerous studies have shown, for example, that humans generally place more value on holding onto an asset they possess than they place on the possibility of obtaining that identical asset in the first place. They similarly broadly assign greater priority to avoiding a loss of a given magnitude than to a gain of exactly the same magnitude.

The subject of risk-taking is fundamental to the creation, construction, and operation of systems that involve humans in their roles as designers of elements of the systems or of the system as a whole. Fortunately, good decisions by humans generally produce good outcomes…but not always. Unfortunately, on occasion, good decisions can produce bad outcomes. Perplexingly, bad decisions can produce good outcomes. With regard to the latter, consider a game of tossing two coins and betting on the outcome: two heads, two tails, or a head and a tail. A bad decision would be to bet on either of the two like-outcomes (two heads or two tails); yet, one-fourth of the time this "bad" decision would produce a good outcome. Similarly, a "good" decision (a mix of heads and tails) would lead to a bad outcome half of the time.

To meet the challenges of systems engineering, its practitioners often depend upon analytical or, occasionally, hybrid models. While of the utmost importance, such models introduce their own perplexities. For example, in modeling a health-care system, what is a human life worth? (One agency of the federal government says the answer is $9.2 million.) Other agencies prescribe the value in terms of something called a "quality-adjusted life-year" (but disagree on its value). There is even the question of defining something as straightforward as what is one's goal. Is it to maximize the likelihood of the best possible outcome? Or is it to maximize the possibility of a good outcome? Or perhaps to avoid the worst possible outcome (i.e., the greatest regret)?

Among the many conclusions, the author of this important book draws the one that stands out pertains to the use of models: "…the best use is as a source of insights, not as a basis for prediction." Wise words, particularly when the use of models is buttressed with sensitivity analyses.

As a society, we are moving increasingly into the realm of nonlinear, interdisciplinary complex systems. This evolution is making us increasingly vulnerable to failings of designers, builders, operators, and users of such systems. Furthermore, in most instances, such shortcomings will be amplified by those who would malevolently exploit weaknesses (read: banking system, energy distribution system, food distribution system, water supply, national security systems, autonomous vehicle networks). Such are the challenges faced by those who would engineer such complex systems. And it is exactly these complexities, and their unintended consequences, that make this book so very useful.

Bethesda, MD, USA

Norman R. Augustine
Retired chairman and CEO of Lockheed Martin,
former under secretary of the Army,
and former chairman of the Council
of the National Academy of Engineering

Preface

This book is about the growing convergence between engineering and other disciplines and how this convergence can be exploited to enable transdisciplinary systems engineering. I define transdisciplinary systems engineering as an integrative thematic discipline that reaches beyond engineering to other disciplines to identify and leverage new concepts and relationships that can potentially make complex system problems tractable and amenable to cross-disciplinary solutions. I envision transdisciplinary systems engineering as a disruptive transformation of systems engineering into a broader, more inclusive discipline capable of addressing scientific and societal problems of national and global significance.

Today, the engineering community acknowledges that new approaches that go beyond traditional engineering are needed to address problems of the twenty-first century. In fact, this recognition provided the impetus for writing this book. Over the last couple of decades, we have seen technology convergence dramatically transform how people live, work, and learn. In this book, I discuss how disciplinary convergence, enabled by technological advances, is already beginning to transform engineering and engineering education. I argue for a new mind-set to address the challenges posed by hyper-connectivity and increasing system complexity.

The twenty-first century is going to be a century of convergence between engineering and other disciplines. We are already beginning to see the potential synergy between engineering and entertainment arts being exploited in systems development, with social media also playing a key role in transforming the engineering landscape. This trend mirrors the synergy that existed between arts and science in a prior era. This synergy was inherent in the thinking and works of renaissance masters like Leonardo da Vinci. Growing specialization over the next few centuries led to the erosion of this synergy. Today we once again stand on the brink of a new renaissance that is being driven by new forms of disciplinary convergence enabled by disruptive technological advances such as 3D printing, Internet of Things, autonomous system-of-systems (SoS), and augmented reality. And for the first time, we are beginning to develop a unified understanding of the physical and cyber worlds at multiple scales.

This new era is characterized by: hyper-connectivity; ubiquitous information access; new forms of communication and collaboration; dynamic, on-demand resource allocation; and uncertain geopolitical futures. There is a greater emphasis on improving human quality of life, the environment, workforce sustainability, and physical and cybersecurity. Process agility and system resilience are rapidly becoming the key sources of competitive advantage. Opportunistic partnering is becoming the norm, and entrepreneurial thinking is becoming increasingly pervasive in both industry and academia. The opportunities presented by these trends inspired me to write this book on how disciplinary convergence can enable transdisciplinary systems engineering and thereby transform both engineering and engineering education.

With ongoing advances in disciplinary convergence, it is time to reimagine the engineering profession without traditional disciplinary boundaries and redefine the engineering professional as one with competencies beyond engineering. These changes require today's engineers to step out of their comfort zones and learn relevant concepts from other disciplines that bear on the engineering of complex, sociotechnical systems. As convergence among disciplines deepens, both academia and industry will have to reorganize and embrace new business models to realize the full benefits of such transformation. This transformation will involve dramatic revisions in engineering curricula and learning approaches.

In this book, I take the reader on an exciting journey that begins with the need to "think different." Along the way, the reader is introduced to disciplinary convergence and transdisciplinary systems engineering which is enabled by disciplinary convergence. Specifically, the reader will learn how transdisciplinary thinking can be exploited in engineering complex systems and system-of-systems (SoS). The journey concludes with a reimagining of the engineering profession and engineering education and identification of promising new technologies that are likely to shape the engineering of complex, sociotechnical systems, and SoS in the future.

This book consists of multiple self-contained chapters that can be read in the order of reader preference. However, to experience this journey as an unfolding story, I recommend reading it in the sequence that the chapters are laid out. This book is well-suited for interdisciplinary courses in sociotechnical systems and SoS engineering. It is also intended to serve as a reference text for individuals interested in pursuing cross-disciplinary research in scientific and societal problems of national and global significance.

Book Organization

In eleven chapters, I tell the story of how hyper-connectivity and disciplinary convergence are becoming the key enablers of transdisciplinary systems engineering. Chapter 1 traces the growth in system complexity as a result of hyper-connectivity and disciplinary convergence. Chapter 2 stresses the need to think different and offers several perspectives where the need to think different has become an imperative. Chapter 3 discusses the key aspects of disciplinary convergence using specific examples and identifies disciplinary convergence as a catalyst and enabler of transdisciplinary systems engineering. Chapter 4 discusses the new face of collaboration with the convergence of distributed collaboration, social networks, and crowdsourcing to spur innovation, facilitate design, and accelerate problem solving. Chapter 5 discusses the convergence of engineering and storytelling, the transformation of models to stories, and the creation of model-based interactive storytelling, a new concept that brings together model-based systems engineering and interactive storytelling to increase and enhance stakeholder participation in upfront engineering. Chapter 6 discusses how psychological principles can be exploited in generating novel options during decision making associated with system architecting and engineering. Chapter 7 brings together various concepts including viewing lenses and system stories within the rubric of experiential design, an interactive engineering approach that enhances stakeholder experience and increases stakeholder participation especially in upfront engineering. Chapter 8 explores the principles of elegant design and discusses how complexity can be effectively reduced and managed by circumventing extraneous system constraints and trade-offs through proper problem framing and structural complexity reduction. Chapter 9 discusses the need for system resilience in the twenty-first century and presents resilience perspectives, along with innovative methods for modeling, analyzing, and designing resilient systems in a way that lends itself to system model verification and testing. Chapter 10 discusses formal deterministic and probabilistic methods that can be employed in the design of autonomous networked SoS. The example of autonomous (self-driving) networked vehicles is used to illustrate the key concepts. Chapter 11 summarizes the key takeaways from the previous ten chapters and examines the future of transdisciplinary systems engineering given the ongoing advances in technology, increasing disciplinary convergence, and ongoing changes in engineering education.

Acknowledgments

Writing a book on transdisciplinary systems engineering was a natural consequence of conducting sponsored research in cross-disciplinary approaches to problem solving and decision making. I was encouraged to write this book by colleagues in academia, government, and industry. While writing this book, my thinking was reinforced by Dean Yannis Yortsos of the University of Southern California's Viterbi School of Engineering, who coined the term "Engineering+" in 2009 to draw attention to the potential synergy between engineering and other disciplines.

My colleagues and collaborators in the aerospace and defense industry were a constant source of inspiration. In particular, Marcus Nance, Allen Adler, John Tracy, Michael Richey, and David Whelan shared their valuable perspectives on the potential impact of disciplinary convergence as a source of competitive advantage in commercial aviation, aerospace, and defense sectors.

I am indebted to Norm Augustine who offered to write the foreword to this book and for his encouragement and support over the years. My thanks to several other friends and colleagues who enthusiastically supported my research and writing of this book. They include John Slaughter, Yannis Yortsos, Jim Moore, Mary Good, Barry Boehm, Bill Rouse, Tom Sheridan, Solomon Golomb, George Friedman, Andy Sage, Ron Sugar, Paul Nielsen, Joel Moses, Randall Steeb, Gerry Estrin, Len Kleinrock, Bill Ballhaus, Wanda Austin, Ann Majchrzak, Dan Erwin, Michael Sievers, Elliot Axelband, John Doyle, Behrokh Khoshnevis, Terry Bahill, Marilee Wheaton, and Ken Cureton. My thanks to my students, Edwin Ordoukhanian, Kurt Klingensmith, and Douglas Orellana. Edwin worked tirelessly to organize the references in this book. Kurt and Douglas provided useful examples to support my arguments.

Michael Luby of Springer was quick to recognize the need for such a book and had my book proposal reviewed promptly. My thanks to him and the other members of the Springer editorial board.

And, finally, I would like to thank my wife, Carla; my sister, Shama; and my daughters, Ayesha, Laila, and Ashna, who offered unwavering support and encouragement throughout the writing of this book.

Los Angeles, CA, USA Azad M. Madni

Contents

Chapter 1
Twenty-First-Century Imperatives

"Today the network of relationships linking the human race to itself and to the rest of the biosphere is so complex that all aspects affect all others to an extraordinary degree. Someone should be studying the whole system, however crudely that has to be done, because no gluing together of partial studies of a complex nonlinear system can give a good idea of the behavior of the whole."

– Murray Gell-Mann, physicist

1.1 Twenty-First-Century Trends

The twenty-first century is already a century of disruptive innovation and ever-increasing complexity that is being fueled by hyper-connectivity and convergence among technologies and disciplines, respectively. According to the 2014 National Academy of Science (NAS) report on convergence, *The key message of convergence, however, is that merging ideas, approaches, and technology from widely diverse fields of knowledge at a high level of integration is one crucial strategy for solving complex problems and addressing complex intellectual questions underlying emerging disciplines.*

Today's products are rapidly becoming smart, connected devices that are increasingly being embedded in larger systems [1]. Specifically, we are seeing a dramatic increase in smart, interconnected industrial systems, smart devices, smart appliances, and smart structures. These systems have three core elements in common: physical subsystems, smart subsystems, and connectors [1]. Physical subsystems comprise mechanical and electronic components. Smart subsystems consist of sensors, microprocessors, storage, controllers, software, embedded OS, and digital user interfaces. Connectors include ports, antennae, protocols, and networks that enable communication between the product and product cloud (which runs on remote servers and contains the product's external OS). The key characteristics of smart, connected devices are:

- *A smart technology infrastructure* that enables data exchange between devices and users and integrates data from business systems, external sources, and related products; this infrastructure also serves as a platform for data storage, analytics, application execution, and device security.

A.M. Madni, *Transdisciplinary Systems Engineering*,
DOI 10.1007/978-3-319-62184-5_1

- *Individual devices* that can monitor and report on their respective conditions as well as that of the environment – this capability helps generate and incorporate previously unavailable insights during device usage.
- *Remotely controlled complex devices* that give users unprecedented ability to customize functions, performance parameters, and device interfaces and operate them in hazardous or hard-to-reach environments.
- *Data monitoring and remote control facilities*, to define routes, set schedules, and allocate resources.

Today smart devices and smart, connected devices are continuing to redefine businesses and industries. They are also reshaping human behavior. The impact and implications of technologies are difficult to understand much less foretell. With unprecedented data volume from smart, connected devices, it has become possible to acquire new insights, which, in turn, can change the way organizations, teams, and individuals interact with each other. Relationships that were once fixed, are becoming dynamic, fluid, and open-ended. The capabilities offered by new devices, the supporting infrastructure, and the resultant data are beginning to reshape every function in the value chain. These include product development, manufacturing, supply chain logistics, and sales and marketing. In addition, the frequency of collaboration among functions is becoming increasingly more important.

1.2 Hyper-connectivity and Internet of Things

Systems in the twenty-first century are gradually and increasingly becoming a combination of cyber, physical, and human components. With the maturation of the Internet of Things (IoT), systems are becoming increasingly hyper-connected and of ever-growing scope and scale. Not surprisingly, the emphasis today is on system and system-of-systems (SoS) affordability, resilience, cyber and physical security, and methods for coping with ever-growing system complexity. With the ubiquity of social media and advent of crowdsourcing, collaboration is being redefined to include a greater number of participants with many unknown participants capable of contributing knowledge, expertise, and partial solutions to specific problems. Advances in low-cost immersive environments, augmented reality, and virtual worlds are making it possible for stakeholders to "experience" an evolving design and concept of operations (CONOPS) firsthand during collaborative development.

Figure 1.1 depicts a complex system along with the factors that contribute to system complexity, methods that mitigate the impact of complexity, and technologies that contribute to the methodologies.

While heterogeneity of subsystems and components increases complexity. Heterogeneity is unavoidable due to the multitude of smart devices, from different manufacturers, that individually and collectively fulfill unique stakeholder objectives. Connectivity among heterogeneous elements also contributes to complexity in that an increase in interfaces and interdependencies among elements leads to expected and unexpected system behaviors, both undesirable and desirable. Hyperconnectivity makes the system vulnerable to systemic failures.

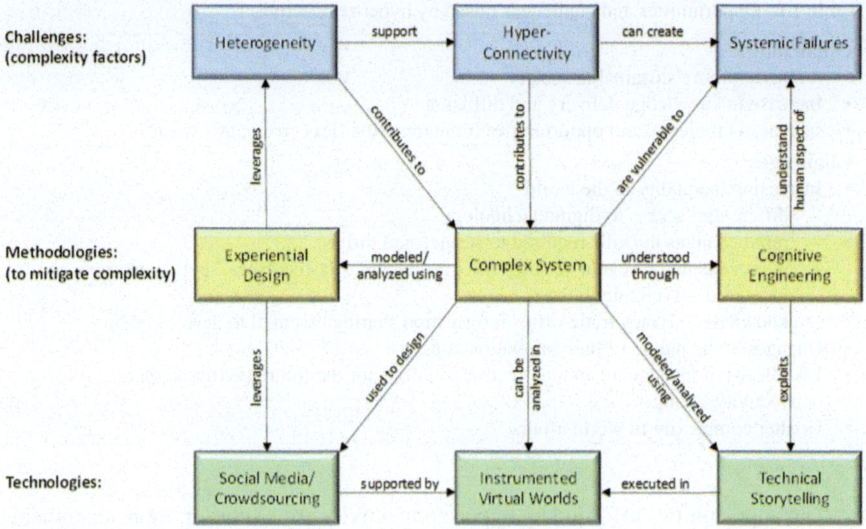

Fig. 1.1 The complexity challenge and technology-enabled methods

While Fig. 1.1 applies to organizations, systems, and networked smart devices. Unique challenges can arise with an increase in networked, smart devices, stemming, in part, from the lack of knowledge of system boundaries. For example, despite its heterogeneity and complexity, large military operations still have some semblance of boundary control. While outside influences exist, military planners are still able to organize their formations with deliberate decisions about what to include and what to exclude depending upon operational objectives. However, in a hyper-connected world, the formation of SoS may be opportunistic and ad hoc and may, in fact, be simply due to the presence of multiple unique, networkable smart devices. Thus, without boundary management, unprecedented SoS may form, adapt, and act with little control. For example, an autonomous vehicle may find itself connecting to a smart structure or appliances while driving through residential areas. And despite boundary control, cyber threats can potentially erode boundary management, and thereby adversely impact our understanding of the behavior of the dynamically evolving SoS. This is the challenge that is currently posed by smart device-driven, network-enabled SoS. It is a challenge that will persist and potentially grow in the foreseeable future.

1.3 Promise and Challenges of Hyper-connectivity

By 2020, it is estimated that 50 billion networked devices will exist in societies and economies [2]. The social, political, and economic implications of this change are only beginning to be understood. To ensure the creation of a human-friendly, hyper-connected world, it is important to understand the nature of the opportunities and the challenges posed by hyper-connectivity (Table 1.1).

Table 1.1 Opportunities and challenges posed by hyper-connectivity

Opportunities
• A smarter, more sustainable society
• Increase in knowledge delivery and diffusion
• Increased efficiency and opportunities for automation (less errors, less waste)
Challenges
• Increasing inequality in the world
– differential access to digital technology
– rapid changes in skills required to survive and thrive
– unprecedented levels of security risks (accidents, "flash crashes," cyberterrorism, and increased surveillance)
– knowledge-privacy trade-offs – information sharing essential to derive benefits
• Changes in the nature of human relationships
• Likelihood of inadequate and ineffective solutions for the future (getting hyper-connectivity wrong)
• Greater complexity in world affairs

The opportunities afforded by hyper-connectivity are a smarter, more sustainable society, increased knowledge delivery and diffusion, and increased efficiency and error/waste reduction through automation. However, the challenges posed by hyper-connectivity are quite daunting. They include increased inequality in the world, changes in the nature of human relationships, and increasing likelihood of inadequate/ineffective solutions for the future. The latter is a direct consequence of getting hyper-connectivity wrong. The greatest and most daunting challenge poised by hyperconnectivity is greater complexity in world affairs.

Today digital technology is becoming increasingly more interwoven into the fabric of society [7]. It touches virtually every aspect of our lives including social interactions, energy, healthcare, political engagement, product development, and everyday decision making. Our reliance on networked resources and hyper-connectivity will continue to grow making us increasingly more interdependent. Hyper-connectivity is already beginning to change the way we live, work, learn, and entertain ourselves.

With ubiquitous access to the web, mobile technologies and social media are making us increasingly more hyper-connected. However, is this a good thing? On the one hand, being in constant contact is clearly transforming our society, especially the younger generation. "Millennials," in particular, are increasingly becoming delegators in the sense that they rely on the internet as their brain [3, 4]. This change in behavior is, in turn, changing the way we frame and solve problems. On the other hand, it can be argued that constant contact is likely to lead to an impatient generation that looks for "quick fixes" to problems regardless of their scale and complexity. Thus, in an era of "constant contact," we need to ask whether complex systems engineering should also be changed? If the answer is in the affirmative, then should these changes be simply allowed to happen, or should they be actively influenced? By incorporating how people work today into the systems engineering process, it may be possible to achieve faster and broader adoption and diffusion of new "best practice processes."

According to the 2012 World Economic Forum report on global risks, critical system failure is a top 50 threat to global development in a hyper-connected world. In this report, the technological risk category identifies the dark side of connectivity with critical system failures as the centers of gravity. Respondents at the Forum viewed risk from the perspective that a single vulnerability could trigger cascading failures of critical infrastructures and networks. They described this type of failure as *low likelihood but high impact*. These types of failures result from combinations of events that invariably tend to catch responders off guard. For example, risks associated with cybercrime and resulting system disruption can dramatically amplify traditional security concerns. Such event cascades can eventually undermine global governance.

In today's world, cyberattacks pose the greatest economic and technological risks that could lead to critical system failures, data fraud/theft, mineral resource supply vulnerability, and a glut of digital disinformation. The report lists five global risk categories: economic, environmental, geopolitical, societal, and technological, each with "an identified center of gravity of the greatest systemic importance, and with critical connectors that link the different categories with respect to Technological Risks." The report states that a critical system failure shares important interconnections with risks associated with the other categories. It is most strongly associated with cyberattacks and three economic risks: prolonged infrastructure neglect, unforeseen negative consequences of regulations, and financial failures of major systems. The report stated that key contributor to technological risk is extreme volatility in energy prices.

The report goes on to say: "The impacts of crime, terrorism and war in the virtual world have yet to equal that in the physical world, but there is a growing recognition that this could change. Online security is now considered a public good, implying an urgent need to encourage greater private sector engagement to reduce the vulnerability of key information technology systems." With cyberattacks identified under the headings of sabotage, espionage, and subversion, the report notes that "national critical infrastructures are increasingly connected to the internet, often using bandwidth leased from private companies, outside of government protection and oversight."

In light of the foregoing, it is safe to say that we are entering uncharted waters. It is up to us to navigate these times with a new, responsible mindset that continually assesses the consequences of hyper-connectivity on our quality of life.

1.4 Promise and Challenges of the Internet of Things

The Internet of Things (IoT), an umbrella term, is beginning to evolve from "connectivity to things" to "data-driven services." The IoT is envisioned as an intelligent network that is sensed and controlled, and can be programmed. It is envisioned that IoT-enabled products will employ embedded technologies to communicate directly/indirectly with each other and the Internet [2, 5, 6].

Technology drivers for the Internet of Things (IoT) are sensors and actuators with their increased interconnectivity to each other and with computing and memory resources. This was entirely expected with electronic devices and sensors becoming increasingly ubiquitous and inexpensive and with device miniaturization continuing at an unprecedented pace. The initial driver for these technologies was the military's need for inexpensive, ubiquitous sensing and computing. For example, DARPA's "smart dust" technology, for the US military in the early 1990s, dates back to the advent of "wearable computers." Today, these technologies have matured to the point where annual meetings in "smart fabrics" and other commercial applications associated with fashion, sports, and medicine have become commonplace [7]. A second driver for ubiquitous sensing and computing are industrial applications and smart manufacturing. These include tracking fleet trucks on the road, detailed mapping of environments, environmental sensing and monitoring using special devices or sensors integrated into smartphones and small tablets, and locating and tracking products in warehouses, transit, and stores. In sports, we have seen instrumentation of both players and the objects they play with (e.g., soccer ball, football). And, there are sensors that are built into sports devices and clothes [7]. Also, ways to reduce restocking costs, using vision processing by mobile robots, have been explored. Related work in datacenter systems attempts to track temperature profiles for improved cooling efficiency. Relatively more recent work aims to find less intrusive ways to monitor or react to command inputs, such as gestures. Games (e.g., Microsoft's Kinect, Nintendo's Wii) have wide market acceptance, underscoring the importance of these technologies for future consumer electronics. Finally, there have been advances in consumer applications such as smart homes, which monitor electricity consumption, adjust to current pricing, and allow owners to remotely access and monitor appliances in their homes. These capabilities appear in the form of smart grids and smart city facilities for security and monitoring. Interestingly, self-monitored oil and gas pipelines to identify pipeline failures predate many of these technologies.

By 2020, the majority of manually performed business processes are expected to be automated as workflows enabled by *embedded sensors* built into products and supply chains and external sensors such as cameras. Smart clothes will enable monitoring of wearers' activities, a useful capability in sports training, the performing arts, and in medical and healthcare settings.

However, as can be expected, there are privacy concerns associated with the IoT [5, 6]. For example, do we want the "cloud" to know our location and activities? The commercial benefits of such instrumentation and monitoring can potentially be offset by the potential for abuse by stalkers and profilers. While monitoring of human activity levels can provide value to doctors, this information can be exploited in harmful ways by others (e.g., by companies that raise an individual's health insurance premiums). Thus, privacy and security concerns will have to be adequately addressed before IoT becomes ubiquitous in our environment [7].

IoT benefits are potentially quite far-reaching from smart infrastructures to smart aircraft. For example, self-monitoring, self-reporting weight-bearing supports for bridges have great payoff. The same is true of smart aircraft wings capable of adjusting their surface structure based on airflow and flight characteristics. Furthermore, the IoT can have a positive impact on the quality of life with smart instrumented cities capable of guiding tourists to historic sites or desired locations. These sites, equipped with smart software (agents), could self-narrate their histories. As important, traffic jams can be avoided by providing real-time guidance in the form of "current best routes." Traffic jams can potentially be circumvented with automobiles moving automatically and responding in consistent fashion to traffic light changes and moving at a common, safe speed. Today, automobile ride-sharing is being automated, and autonomous cars are becoming capable of arranging rendezvous points and describing points of interest along the way.

In fact, the IoT, with its ubiquitous sensor awareness, affords new opportunities that can dramatically enrich our lives, while making the instrumented world into an exciting research test-bed. In one particular demonstration of the power of the IoT, a phone company tracked its cellphones' current locations, to draw a precise map of London, including boats in the river. This example demonstrates the impressive progress in mapping technologies made possible by the IoT [7].

The foregoing examples clearly show that the IoT is more than a mere collection of distributed, embedded sensors and actuators. It is augmented, enabled by, and dependent on distributed data collection and computing resources. Its growth and evolution depend largely on the ability to deploy the computational and storage capacities needed to support its growth. That, in turn, depends on the potential benefits of building and paying for such infrastructure. However, there are potential hurdles. First, it is important to identify which entities will benefit and, therefore, will be willing to make ongoing investments. Currently, large corporations such as Google are capitalizing on end-user services, but there are other models that manifest themselves through the use of "micro-services" as in banking via cellphones in developing countries. Clearly, if such developments become more ubiquitous, they will reduce reliance on large companies. Second, as with peer-to-peer technology, there are legal and governance issues as well which, if left unresolved, can impede progress. A notable legal case in the United States was made against an individual wearing Google Glass and driving a car. He ended up with a traffic fine because of the laws passed against texting while driving [7].

Ultimately, the IoT is driven by advances in device technologies. Unprecedented opportunities are being created by billions of smartphones with built-in sensors, internet connectivity to fixed facilities, and mobile connectivity. The new capabilities enabled by IoT directly support business goals (e.g., reduced cost through automation, shorter cycle times for supply chains).

1.5 Complexity Challenges

An increase in the number of interconnections and interdependences among system components is a source of ever-increasing system complexity. At the same time, advances in computing and communication are creating a steady shift toward cyber-physical human systems. Along with this trend comes the need for precise control of timing to assure safety and correctness [7]. This trend poses a challenge to system modeling, analysis, and complexity management because the intellectual and technological basis of current approaches tend to be rooted in models and tools, including APIs, high-level languages, and virtualization that abstract away time! Therefore, there is a pressing need to develop new, holistic approaches to model, analyze, influence, and shape complex systems behavior. As important, there is a need to develop new approaches to manage complex sociotechnical systems and their evolution. A reasonable starting point for understanding complexity is to consider at least five properties that are common to natural, biological, and social systems: nonlinearity, scale dependence, recursive symmetries between scale levels, sensitivity to initial conditions, emergence resulting from multiple feedback loops, and interaction chains [8]. Engineered complex systems exhibit other key properties such as drift into failure and interacting with the protective structure that is put in place to prevent them from failing.

1.6 Summary

The twenty-first century is a century of exponential growth in complexity resulting from hyper-connectivity. It is also a century of disruptive innovation fueled by increasing media integration and disciplinary convergence. We are already seeing a dramatic increase in smart, interconnected devices and systems that are supported by smart technology infrastructures and data monitoring and remote control facilities. These advances are redefining entire businesses and industries (e.g., healthcare, manufacturing).

Systems in the twenty-first century are increasingly becoming a combination of cyber, physical, and human components. These systems are expected to have the capacity to learn and are required to be trusted and trustworthy. As systems learn from data, software, and people, people have to adapt as well to stay in harmony with systems that continually adapt through learning. Learning, and more specifically machine learning, can be a source of competitive strategy and a source of enduring competitive advantage.

In sum, hyper-connectivity offers several advantages even as it creates unprecedented challenges. The advantages are an increase in knowledge delivery and diffusion and customization of functions that result in fewer errors and less waste. The negative consequences of hyper-connectivity are differential access to digital technology contributing to growing inequality in the world, dramatic increase in complexity in all walks of life, and ever-growing security risks, and privacy challenges.

Today digital technology is becoming increasingly and inexorably interwoven into the fabric of society. With 24/7 access to the web, mobile technologies, and social media, human behavior is changing in unexpected ways. Millennials, in particular, are increasingly becoming delegators and multitaskers with short attention spans. These characteristics are breeding impatience and the search for "quick fixes."

As important, these characteristics are influencing the way problems get framed and how material is taught in the classroom. In many ways, we are entering an era of the unexpected and unanticipated.

In light of the foregoing, we need to learn how to exploit hyper-connectivity and convergence to accelerate work processes and manage complexity. In particular, the convergence of engineering with other disciplines will be enabled by technological advances. Convergence between engineering and other disciplines can take a variety of forms: engineering makes discipline X more efficient, engineering empowers or enhances X, the combination of engineering and X leads to a new field/discipline, and engineering and X jointly solve an apparently intractable problem. Ultimately, these various types of convergence will become the enablers of transdisciplinary systems engineering, the central theme of this book [9].

References

1. Porter ME, Heppelmann JE. How smart, connected products are transforming competition. Harv Bus Rev. 2014;92(11):64–88.
2. Convergence: Facilitating Transdisciplinary Integration of Life Sciences, Physical Sciences, Engineering, and Beyond. National Academy of Science, Committee on Key Challenges areas for Convergence and Health; board on Life Sciences; Division on Earth and Life Studies, National Research Council, 2014.
3. Greenemeier L. Piece of Mind: is the internet replacing our ability to remember? 2011. [Available from: http://www.scientificamerican.com/article/internet-transactive-memory/.
4. Harris M. How the web became our "external brain," and what it means for our kids 2014 [Available from: http://www.wired.com/2014/08/end-of-absence-how-technology-changes-our-brains-society-children/.
5. Porter ME, Heppelmann JE. How smart, connected products are transforming competition. Harv Bus Rev. 2016;92(11):18.
6. Porter ME, Heppelmann JE. How smart, connected products are transforming companies. Harv Bus Rev. 2015;93(10):53–71.
7. Alkhatib H, Faraboschi P, Frachtenberg E, Kasahara H, Lange D, Laplante P, et al. IEEE CS 2022 report. IEEE Comput Soc. 2014;25–27.
8. Roco MC, Bainbridge WS. Converging technologies for improving human performance: nano-technology, biotechnology, information technology and cognitive science. Norwell: Kluwer Academic Publishers; 2003.
9. Madni AM. Transdisciplinarity: reaching beyond disciplines to find connections. J Integr Des Process Sci. 2007;11(1):1–11.

Chapter 2
Thinking Different

"The world as we have created it is a process of our thinking. It cannot be changed without changing our thinking."

– Albert Einstein, theoretical physicist

2.1 Systems in the Twenty-First Century

The twenty-first century will continue to be defined by disruptive innovation, hyper-connectivity, and increasing scale and complexity of systems and system-of-systems (SoS). The twenty-first century will also be an era in which autonomous systems and system-of-systems (SoS) will become increasingly prevalent. These advances will surface new concerns rooted in security issues, ethics, and law. These new concerns are beginning to surface new trade-offs and challenges for society to address. Collectively, addressing these concerns will require new types of thinking that emphasize trade-offs, cybersecurity, flexibility, adaptability, and resilience, rather than point optimizations.

The new types of thinking needed today coupled with increasing convergence among disciplines are redefining what we mean by literacy in the 21st century. According to Alvin Toffler, *"the illiterate of the 21st century will not be those who cannot read and write, but those who cannot learn, unlearn, and relearn."* Today's mindset, the legacy of several past generations, will need to change because many historical assumptions no longer hold. For example, the current mindset views the world as (a) *deterministic* and characterized by clarity and confidence, "cause and effect," and "a single best answer" or (b) *probabilistic*, that is, characterized by uncertainty but predictable within limits. This view, rooted in the twentieth-century mindset of quantum and statistical theories [1], assumes the world is probabilistic with bounded predictability. People who espouse this world view understand concepts such as variations and variances, but not how change propagates in complex systems [36]. The twenty-first century is characterized by hyper-connectivity, complexity, uncertainty, ambiguity, and cyber-threats. It is also a century in which convergence will fuel disruptive innovation. Being successful at innovating in the twenty-first century requires the ability to envision plausible, alternate futures in different time frames including the time frame of interest. Integrating the future with the present is also a challenge because of the difficulty in envisioning alternate

A.M. Madni, *Transdisciplinary Systems Engineering*,
DOI 10.1007/978-3-319-62184-5_2

futures with today's mindset. Envisioning the future requires a totally new mindset that employs new ways of thinking to make sense of the world, understand and manage complexities, and exploit opportunities.

2.2 New Types of Thinking

The impetus for thinking different in the twenty-first century stems from technological advances, changing human roles and behaviors, poorly understood cyberthreats, need for increased rigor in modeling and analysis, and the need to be opportunistic. In the following paragraphs, I discuss the different types of thinking that we need in the twenty-first century.

2.2.1 *Thinking About Thinking*

Twenty-first-century engineers need a big picture mentality and the ability to think in terms of interconnectedness of various elements that form larger systems. This type of thinking, called holistic thinking, is in sharp contrast to reductionist thinking found in traditional systems engineering. Reductionist thinking employs analytic reasoning and is concerned with breaking down a system into its constituent elements, analyzing the elements, and composing the solution to the system from the solutions to the elements. Holistic thinking, which exploits systems thinking, is concerned with understanding how elements within a system influence one another and how the whole system interacts with its environment. Systems thinking also addresses the issue of defining system boundaries, a difficult proposition that varies with context. Specifically, systems thinking is rooted in asking "smart questions" to identify the variables that need to be included within the system boundary. It is important to realize that as more detailed questions are posed, the boundary may have to be expanded to include additional entities within the system boundary. This is because the initial system boundary will invariably need to be revised as interactions of the system with the assumed environment are better understood. With the understanding afforded by systems thinking, system architects and engineers will be able to identify leverage points that will increase the odds of leading to desired outcomes. Systems thinking needs to be accompanied by critical thinking, which is rational thinking with the ability to offer critique and engage in debate.

2.2.2 *Thinking About Humans*

System complexity, technological advances, and operational demands are redefining the roles of humans in relation to systems. The role of the human is shifting from an operator of a system to an agent within the system [2, 3]. As important, the interactions between humans and systems are becoming increasingly more complicated as humans are required to interact with adaptable systems. And then there are lingering misconceptions about humans that adversely impact system architecting and human-system integration. Perhaps the single biggest misconception is that humans are "suboptimal job performers" who need to be shored up and compensated for during joint task performance. This view of humans as liabilities fails to capitalize on human strengths such as creativity and ingenuity [2, 3]. Also, the resulting system architecture and human-system integration approach forces humans to work with the system in a manner that is inherently incompatible with human conceptualization of work. Today it is abundantly clear that a balanced view of humans is called for – one that capitalizes on human capabilities while circumventing human limitations [2, 3]. This view recognizes that while humans are capable of adapting to changes in context, such adaptivity is neither complete nor fast. In fact, frequent context-switching adversely affects human performance. Similarly, both infrequent and overly frequent stimuli also degrade human performance. With infrequent stimulus, human vigilance drops, resulting in high likelihood of missed detection. With overly frequent stimulus, humans are unable to keep up with the stimulus arrival rate resulting once again in missed detection and degraded performance.

2.2.3 Thinking About Models

Models are increasingly being used to structure and organize knowledge about a system. Models also offer a means to reduce structural complexity and acquire insights into system behavior under a wide variety of assumptions and conditions. As important, models help contextualize knowledge and facilitate collaboration. Models, by definition, are abstractions and approximations, and therefore do not fully reflect real-world systems. Often, models are difficult to verify and test completely. Therefore, their best use is as a source of insights, not as a basis for predictions. The question according to Box and Draper [4] is "how far can models be wrong before they cease to be useful." This is a useful view when discussing models. Today, models are the centerpiece of model-based systems engineering (MBSE). They serve to replace the document-centric view of systems engineering with a model-centric view from which documents can be automatically generated on demand.

The information needed to model a system is seldom available in full. Therefore, assumptions and educated guesses have to be made during model construction. However, as new information becomes available, these assumptions can be revised or replaced with actual evidence (i.e., data). It is important to realize that assumptions have to be made when actual data is not available. Assumptions made can be warranted or unwarranted. It is important to ensure that unwarranted assumptions are avoided when modeling the system and the operational environment.

2.2.4 Thinking About Trade-offs

People perform trade-offs instinctively every day without ever thinking about them, because everyday trade-offs tend to be simple, straightforward and well-known. However, trade-offs can become quite complicated in disciplines as different as engineering design and sports. Often, they require human judgment and invariably depend on evidence, i.e., data. In the following paragraphs, I review trade-offs in popular sports (e.g., baseball, American football, professional basketball, and motorsports) to convey their subtlety and communicate their importance. When it comes to engineering, these tradeoffs are much more complicated and not always apparent. As important, problem framing impacts the tradeoff space.

Baseball Baseball is a trade-off rich game. One trade-off that is easily understood is the decision to bunt in specific game contexts to advance a baserunner while risking the bunter being thrown out at first base. The trade-off here is between (a) choosing one of three total outs per inning and (b) moving a baserunner closer to a scoring position thereby increasing the likelihood of scoring. *The key concept here is to deliberately suboptimize locally for a potential long-term gain.*

American Football (1): Fourth Down Decision In the National Football League (NFL), teams get four opportunities (downs) to move the ball 10 yards. If they are unable to, they have to turn the ball over to the opposing team. The "four-downs rule" is a constraint that drives the fourth down decision. The most appropriate (i.e., potentially effective) decision on fourth down depends on context. The choices available to the team on offense (e.g., with the ball) are to punt, attempt a field goal, or "go for it." Each option has pros and cons. The decision to punt trades possession (increases risk of loss) in favor of the opposing team's potentially poor field position (decreases risk of loss). This is because by punting, the team with the ball gives up possession of the ball, but an effective punt can bury the opposing team deep in its own territory, thereby mitigating the risk associated with giving up (i.e., surrendering) the ball. The decision to punt makes eminent sense if the team with the ball is facing fourth down in its own territory or well outside the opposing team's territory. Tedy Bruschi, a retired New England Patriots' football player, makes a persuasive argument that effective punting can play into "complementary football" and produce hidden offensive gains. *The key point here is that pursuing a defensive course of action can produce hidden gains.*

A *field goal* trades ball possession for an opportunity to score 3 points while also having the likelihood to pin the opposing team deep in its own territory in a poor field position, assuming the field goal is successful and the following kickoff pins the opposing team deep in its territory. However, there is always a chance that the field goal misses even though today's NFL kickers are incredibly reliable, with some being almost "automatic." Nevertheless, making a field goal is not a sure thing. The risks and benefits of kicking a field goal vary as a function of kicking distance, score, and time. For example, if the team with the ball is down by 4 or more points with under a minute to go, opting to kick a field goal is a poor choice.

On the other hand, if the team with the ball is leading comfortably and facing fourth down at the 35-yard line, then deciding to kick a field goal is a lower-risk option with potentially reasonable payoff (i.e., padding the lead by 3 points).

Going for it on fourth down trades the likelihood of turning the ball over to the opposing teams if they cannot convert fourth down, for an opportunity to retain possession of the ball and control the clock. The risk of failing to convert fourth down varies with field position and can be significant. For example, failing to convert in the red zone risks loss of a field goal opportunity. On the other hand, failing to convert at midfield or in one's own territory gives the opposing team great field position. Needless to say, converting on fourth down is a high-payoff event. It allows the team with the ball to maintain possession while affording the team multiple alternatives on offense. Furthermore, converting on fourth down gives a psychological boost to the offense. "We converted on fourth down! We're unstoppable! Our game plan is bold! Our coach knows what he's doing!" The psychological effect on the defense is exactly the opposite. "We had them where we wanted but could not stop them! They converted on fourth down! We're losing control of the game! We're tired! We thought we were getting off the field, but now, we have to stay on and defend again!" According to Gregg Easterbrook, a sportswriter, there is a sociopolitical dimension to fourth down trade-offs. Coaches generally want to do the safe thing, which is to follow conventional wisdom, i.e., on fourth down, send in the kicking unit. Then if the team loses, poor player execution is blamed for the loss. On the other hand, if the coach decides to go for it on fourth down and the team fails to convert, the coach is blamed for having made "a strategic blunder." Thus, in following conventional wisdom, the coach trades higher-risk, higher-payoff outcome for social and political capital. *The key concept here is that social perception and politics sometimes drive decisions to pursue the safe option.*

There is, however, a cultural aspect to this trade-off as well. It has been suggested by some sportswriters that losing college coaches tend to opt for meaningless field goals to narrow the margin of defeat as opposed to going for it. This choice impacts "style point" when voting to rank teams and can also have sociopolitical benefits when the college administration reviews the accomplishments of the coach. "I lost to the top-ranked team by only 11 points!" This is dramatically different from the NFL, where a loss is a loss, regardless of style or margin.

According to Football Outsiders, most teams seem to approach offense in football with the mindset of having only three downs to cover 10 yards, with the fourth down play always being a kick (field goal or punt). This means that the offense must average 3.33+ yards per play versus 2.5, which in turn impacts play calling and schemes. In this case, the conventional wisdom is to kick on fourth down, and the preconceived notion is that there are only three downs to cover 10 yards, not four, and the conventional wisdom is to kick on fourth down. *The key concept in this example is that preconceived notions and conventional wisdom can limit the available courses of action.*

American Football (2): Defensive and Offensive Schemes Unless a defense or offense is exceptionally balanced with top players, it is highly likely that each unit will have strengths and weaknesses. This fact results in schematic trade-offs for game plans. For example, if a team has a strong running game, but a weak passing game with a mediocre quarterback, then the opposing team's defense may invest completely in stopping the run. This implies stacking the line at the expense of the defensive backfield. *The trade-off here is to invest in stopping/containing the effective tactic while forcing the offense to go with the second or third play choice.*

There are also situation-specific schematic trade-offs. For example, on a third down, long-yardage situation, there is a high probability that the offense will pass. Therefore, the defense is likely to decrease the number of players at the line while dropping several players back for pass coverage. The trade-off here is the defense accepting the risk of giving up yardage to the run while decreasing the likelihood of passing yardage, which can more easily and quickly make the necessary yardage for the first down. *This is an example of focusing on stopping the most dangerous (or lethal) course of action while deliberately increasing the occurrence likelihood of less dangerous courses of action.*

And then there are unconventional trade-offs. In the New York Giants and New England Patriots Super Bowl 2012, NE let NY score. Alternative 1 was to give up the touchdown and regain possession down by 4 points with 57 s to go and two time-outs remaining. Alternative 2 was to hold NY to a field goal, going down by only 1, and have 20 s or less to drive 50 yards for a field goal (with no time-outs remaining because NE would have to use them on the defensive stops). Coach Belichick concluded that *time* was more valuable than *points* and let NY score. While sole reliance on analytics may suggest a particular alternative, a great coach will tend to rely on more than analytics to make the call. For example, in big, emotional games like the Super Bowl, intuition and perception based on factors such as momentum play a significant role in coaching decisions. *The key concept here is that in certain emotionally charged contexts, a decision based on perception of momentum and intuition can trump one based solely on analytics.*

Another example of intuition and perception over analytics and conventional wisdom can be found in the 2015 New England Patriots' Super Bowl victory over the Seattle Seahawks. The game came down to a series of goal line plays. In this case, conventional wisdom suggests that the defending team should call time-out to preserve time, which then creates a future opportunity to possess the ball. Coach Belichick thought otherwise. He let the sequence play out and in so doing surrendered the opportunity for a future offensive possession. He gave this up deliberately to maintain momentum and retain the prevailing dynamic of the game. Something didn't feel right to him. He was right. His defense made the stop. Intuition, perception, and momentum won out.

Basketball The final minutes of a close National Basketball Association (NBA) game often devolve into a fouling competition, in which the trailing team intentionally and repeatedly fouls the other team. Each foul stops the clock, affording the leading team the opportunity to make two free throws. The trailing team then gets possession of the ball with the hope of scoring and cutting into the lead. So, what is the trade-off here? The trailing team allows the leading team to potentially extend their lead (by making one or two free throws) in exchange for stopping the clock and regaining possession. The opportunity cost of stopping the clock and getting the ball back outweighs the threat of the leading team extending its lead. It is important to realize that in this case, the trailing team may never get another opportunity to tie or win a close game without resorting to fouling a poor foul shooter on the leading team.

There are layers to this strategy. A smart team will employ analytics (prior to the game) and a priori target a poor foul shooter on the opposing team. This strategy reduces the likelihood of giving up points. The "Hack-a-Shaq" strategy employed against Laker great Shaquille O'Neal in his playing days is an instance of this strategy. There are other factors that bear on this decision such as psychological or cognitive disruption caused by interrupting game flow. And, finally, there is emotional distress and frustration that results from being fouled repeatedly. The latter is a means to "get inside a player's head" causing him to miss.

Motorsports A common strategic trade-off encountered in NASCAR racing occurs during (what is supposed to be) the final pit stop (service stop) of the race. The final pit stop is usually determined by various factors that go into defining context (e.g., what are the other teams doing, is the race under green flag or caution and dynamic constraints (e.g., fuel remaining, tire degradation)). Team crew chiefs invariably employ their perceptual, analytical, and cognitive abilities to make an educated guess about when the final pit stop is likely to occur. At that time, the available options are (1) do not pit, (2) fuel only, (3) fuel and right side tires (changing two right side tires is effective in oval racing), and (4) fuel and four tires. The factors that influence this decision are fuel range, pit stop time, and potential performance gain. Table 2.1 summarizes these options.

The trade-offs and attributes become especially interesting in race context. Each team is its own agent, with team decisions ranging from nearly uniform to highly random. While "conventional wisdom" might suggest a particular alternative, it is not uncommon for some teams to gamble on other alternatives. The time remaining in the race also has an impact. Specifically, is there sufficient time left in the race to exploit the potential performance gain from new tires to overcome the time and

Table 2.1 Pit stop options and associated trade-offs

Do Not Pit	**Pros**: Zero pit stop time, maintain track position, or improve it as others in front pit **Cons**: Risk of running out of fuel, no performance gain of fresh tires
Fuel Only	**Pros**: Lowest pit stop time, eliminates risk of running out of fuel **Cons**: No performance gain of fresh tires
Fuel and Right Side Tires	**Pros**: Eliminates risk of running out of fuel, moderate performance gain **Cons**: Increased pit stop time, greater loss of track position without maximizing performance gain
Fuel and Four Tires	**Pros**: Eliminates risk of running out of fuel, highest performance gain **Cons**: Longest pit stop time, most track position lost

position lost from pit stop? Individual car performance is a key factor. A poorly performing car may choose to skip a pit stop knowing that while it won't win, it may finish better than its current position.

2.2.5 *Thinking About Change Propagation*

A key aspect of the new mindset is understanding how change propagates in complex systems. The intent in this case is to understand the downstream impact of a "quick fix" that might solve an immediate problem only to surface later in the form of unintended consequences elsewhere in the system and displaced in time. The following example presents an example of change propagation.

An unmanned aerial vehicle (UAV) equipped with multiple sensors employs a sensor fusion algorithm to locate objects of interest. The onboard fusion algorithm is slow, so information is generated slower than real time. To speed up the process, the engineering team recommended incorporating a faster processor for sensor fusion and to improve communication capability. This concept is adopted and the sensor fusion latency problem is solved. However, it is discovered that the faster processor generates excessive heat that needs to be dissipated. The engineering team proposes a cooling unit to be housed onboard. This proposal is accepted and the cooling unit is incorporated onboard. The introduction of the cooling unit alters the dynamics of the UAV and reduces its payload carrying capacity. The change in dynamics and the center of gravity may turn out to be acceptable or may destabilize the UAV causing it to crash.

In the preceding example, we viewed the UAV in terms of its subsystems and potential instability that might arise from making a change to one of the subsystems. However, we can view the UAV holistically from an "enterprise" perspective. From this perspective, the introduction of new component(s) may produce instability through hidden avenues and interactions. For example, the logistics support system for the UAV may now need to provide more parts and equipment, introducing complications to the supply line. Also, maintenance personnel, trained on the old UAV configuration, may no longer know how to work with and on the new system. Furthermore, while the new UAV may offer improved communications, the supporting network might not have adequate bandwidth or coverage to exploit the new system. Thus, while the end system may function properly, the supporting and enabling systems may be inadequate potentially introducing instability in the UAV enterprise.

A real-world example of the above problem was brought to my attention by my student serving in the US Army. His infantry platoon had both RG31 MRAPs and HMMWVs in Afghanistan. The RG31 MRAP was intended to fulfill the tactical ground vehicle role that was previously filled by the HMMWV. In addition to other improvements, the RG31 MRAP's main benefit was that it provided superior protection from IEDs. Despite being a better end product overall, the MRAP system as a whole was "unstable" because its logistics support system did not carry large stores of available parts and the mechanic assigned to their outpost was neither trained nor equipped to perform full maintenance on the RG31 MRAP. Furthermore, a few part numbers for the RG31 MRAP were not in the Army's maintenance system. Additionally, only certain operating bases had contracted workers who were certified and equipped to fix a MRAP. Given the harsh environment and duty cycles, broken and worn parts were inevitable for any vehicle. However, a broken leaf spring on the RG31 MRAP, which under ideal conditions could be fixed in 1 to 2 days, instead took months to repair, with a large fraction of lost time resulting from the vehicle having to sit in queue waiting for a part to arrive. This delay introduced "instability" into an otherwise excellent tactical vehicle platform, resulting in patrols that could not always use the RG31 MRAP. This instability meant that they lost the survivability (and psychological) advantage provided by the RG31 MRAP.

Fortunately, with humans in the loop, they were able to adapt and work through their problems. Parts scavenging from disabled vehicles helped to some degree. Maintenance issues were resolved by either authorizing untrained mechanics to work through the problem as best as they could or by making a trade-off foregoing an operational patrol in favor of traveling into a base with trained maintenance personnel. However, these inelegant "holdover" solutions were far from ideal, often creating additional problems that had to be dealt with in the future (e.g., trucks that would normally need only one fix ended up needing more because their good parts were stripped, they had to fix incorrect repairs, or there were operational delays due to maintenance).

2.2.6 *Thinking Wicked*

Most complex problems today are sociotechnical problems, associated with sociotechnical systems. These problems tend to be "wicked," in the sense that they are resistant to solution. So, what makes a problem resistant to solution? Table 2.2 presents some of the salient characteristics of a wicked problem.

From Table 2.2, it is apparent that attacking wicked problems requires a different mindset. In particular, asking the right questions and understanding interdependencies and change propagation paths become important. For example, the question, "what causes traffic congestion?" seems rather straightforward at first blush. Too many automobiles on the roads seems like a reasonable answer. But, is it? Hardly. Weather, ongoing construction, traffic accidents, or cargo that has fallen off transport vehicles are all potential sources of congestions. What about work locations (industries) and residential areas? What about locations of shopping malls and amusement parks? And, of course, throw in the cost of gas and travel time. And, then we have the various ways in which change can propagate in this complex system. Consider also the fact that roads are added to alleviate traffic congestion but in fact can become the cause of greater traffic. For example, reduction of traffic congestion can shorten travel time from source to destination, which, in turn, changes the behavior of people – they begin to commute more than before causing unexpected congestion.

Table 2.2 Characteristics of wicked problems

• No definitive description or formulation
• Always a symptom of another problem
• Tend to be unique
• Always have multiple explanations – appropriateness of explanation depends greatly on perspectives and values of interested stakeholders
• Solutions can be characterized as only good or bad, not right or wrong (true or false) – solutions tend to be largely a matter of judgment
• Cause and effect tend to be elusive – nonlinear, multiple interdependent variables resulting in multiple feedback loops – can be displaced in time and space (delays and distances)
• Multi-objective – overall system adapts – local agents (individuals) optimize locally
• Sensitive to initial conditions (history)
• No surefire prescription to follow
• Solutions can generate unexpected and unintended consequences over time, making it difficult to measure their effectiveness

The foregoing realization leads to a quandary. How should this problem be framed, and what is the desired state of the complex system? Remember traffic congestion affects us in a variety of ways: increased pollution, potentially increased crime, and transportation bottlenecks. Then again, growing traffic congestion can be a symptom. It could be a symptom of unexpected employer policy or unexpected human behavior (e.g., lack of acceptance of telecommuting). It could be a symptom of population growth and urbanization. It could also be a symptom of inadequate mass transit systems.

The agents within the complex system will pursue local optimization (i.e., what is best for them). This implies that they may change their commute routes, jobs, and homes based on their perception of the situation and the things they value. And, of course, wicked systems/problems are complex systems and, therefore, sensitive to initial conditions (e.g., location of residential areas and industries, speed limits, traffic accidents yet to be cleared, routine traffic patterns, and prevailing perceptions about the impact of autos and transportation network). So, back to the original question – in the absence of a clear, noncontradictory definition of the problem, where does one start? The answer is to have realistic expectations of what is achievable and what is not with the solutions proposed to wicked problems. First and foremost, one can never understand all the intricacies of a wicked problem. In fact, the very act of investigating the problem can change the problem itself. Wicked problems, viewed as complex systems, cannot be controlled in the traditional sense. This is because the more precisely you want to define the problem, the less accurate you are likely to be. Thus, definition of a wicked problem eliminates other plausible definitions and has the unintended effect of artificially restricting the solution space, and thereby the set of acceptable options. A wicked problem at any point in time is unique and is likely to morph over time. From the foregoing, one can reasonably conclude that for wicked problems, outcomes (i.e., alternate futures) can be anticipated to a lesser and lesser degree with increasing time horizon, and never predicted with any degree of confidence.

So, one is compelled to ask – where do we start to "gain a grip" on a wicked problem. I use the phrase "gaining a grip" as opposed to "solving the problem" because wicked problems are not amenable to finding a solution in the traditional sense. Thus, given that attempting to define the problem is futile, one needs to begin by identifying different formulations of the wicked problem, so that each can be independently explored to gain insights. To develop a better understanding, we need to: role-play different agents (i.e., participants) in the wicked problem to understand the problem from different perspectives; focus on the ability to rapidly adapt and adjust to new insights, events, and conditions, rather than think about optimization; continually collect and analyze data to discern patterns and the "stories" resident in the data; continue to gain additional insights, hypothesize, and test hypothesis using a variety of gaming, simulation, social network analysis, and data analytics tools; continue to gather information on quantities of interest, revalidate assumptions, and postulate and test new hypothesis (remember that quantities of interest typically change with context, so ensure you have the requisite flexibility in data collection); strive to uncover underlying structures and patterns while addressing the problem in different time frames (near-, mid-, and long-term); continue to adapt to be in a preferred state defined by dynamic stability in the face of limited chaos; finally, never lose sight of the relationships between humans/society and technology, and humans/society and the environment to stay grounded in the real world.

In light of the foregoing, my recommendations for engineering complex sociotechnical systems are to start with questions, not requirements, develop multiple formulations of the problem to investigate and glean insights, interact with different problem formulations to gain insights about the sociotechnical system under study, identify technology levers and impacts of technology on humans and societies, and develop models and tools that are designed to evolve with use in unimagined ways as the wicked problem is studied from different perspectives. This strategy is intended to allow the researcher/engineer/policy maker to understand the problem, not attempt to solve it. Gaining a grip means understanding interventions that can be applied to maintain the wicked problem (a complex system), in a state of dynamic equilibrium in the face of some chaos.

In sum, "wicked" refers to the lack of a clear and fixed problem in which complexity resides in both the problem and solution spaces. Social and sociotechnical complexity play a significant role in wicked problems. Current tools are woefully inadequate for addressing wicked problems. The key problem with current tools is they require crisp information as inputs. However, design of systems to address wicked problems suffers from both epistemic uncertainty (uncertainty due to lack of knowledge) and aleatory uncertainty (uncertainty due to uncontrollable randomness). As a result, judgment, possibility, probability, preference, and other soft factors become the predominant factors in decision making. This means that decision-relevant information may be crisper, imprecise, or any combination of the two. With wicked problems, the emphasis should be on "getting a grip" on the state of the system rather than trying to solve a problem because wicked problems continue to morph over time and in response to certain stimulus/phenomena. Therefore, the emphasis should be on containing the problem to prevent unintended consequences rather than on making predictions based on an accurate model of a system that is not possible with wicked problems.

2.2.7 Thinking Complexity

Twenty-first-century engineers will need a big picture mentality and the ability to recognize the interconnectedness of various elements that form larger systems, patterns, and objects. This type of thinking is referred to as holistic thinking. Thinking holistically is the opposite of thinking analytically, i.e., breaking down a system into its component elements. Systems thinking is a holistic approach that is concerned with understanding how elements within a system influence one another and how the whole system interacts with its environment. With this understanding, humans are able to identify leverage points that lead to desired outcomes. Critical thinking is clear-eyed, rational thinking with the ability to offer critique. Systems thinking employs various forms of thinking including analytical thinking and associative thinking. Analogical thinking is what we do when we use information from one domain (e.g., biology) to help solve a problem in another domain (e.g., cybersecurity). Experts often use analogies during problem solving. In fact, analogies have been exploited in numerous scientific discoveries. Associative thinking is the process of linking between a particular subject with all pertinent factors without drawing on past experience.

When it comes to complex engineered systems, several of these types of thinking come into play. These require a fundamental shift in mindset. Table 2.3 presents the change in mindset that is needed when it comes to system design and operation.

Systems engineers attempt to understand and design a system by modeling the envisioned system, decomposing the system model into component elements, analyzing each component to create a designed solution for each element, and integrating (composing) the designed solutions for the different elements into an overall system design. This hierarchical decomposition approach, based on reductionism, works for well-defined systems with specified boundaries. Airplanes, automobiles, phones, and highways are examples of such systems. Traditional systems engineering depends on being able to decompose a system into system elements and recompose solutions for each element and, as important, having a responsible agent (e.g., human) with the authority and resources to design and build the system. However, this approach is inadequate for complex engineered systems. Specifically, decomposition may produce information loss because of interactions among system

Table 2.3 New thinking needed for complex engineered systems

• Holism, not reductionism
• Self-organization, not organization design
• Agility and resilience, not optimization
• Scaling up methods, not dumbing down problems
• Interdisciplinary, not disciplinary
• Influence, not control
• Incentives and inhibitors, not command and control
• Mutual commitments, not contractual relationships
• Adaptation and learning, not pre-scripted response

Table 2.4 Complex engineered systems characteristics

• **Nonlinear dynamics** – Random, chaotic behavior; does not reach equilibrium
• **Independent agents** – Physical, psychological, social; rule based
• **Conflicting goals, behaviors** – Heterogeneous needs/desires/beliefs reflected in how agents adapt to each other's behaviors
• **Intelligent agents** – Can exist in real/virtual world; overall behavior changes over time
• **Adaptation and learning** – Self-organization (emergent behavior patterns); emergent behaviors may be in the form of valuable new information or unfortunate accidents
• **Influencing behavior** – No single point of control; behaviors unpredictable and uncontrollable; no one in charge; behavior more easily influenced than controlled

elements that potentially produce emergent phenomena. A second problem is that no agent is in charge with the authority and resources to design the system. These are two significant limitations of complex engineered systems. Table 2.4 presents the key characteristics of complex engineered systems.

There are several implications that stem from these properties of complex engineered systems. Most important, these systems cannot be controlled in the traditional sense, i.e., made to comply with behavioral and performance requirements. Agents within a complex engineered system can have adequate smarts to act in their own self-interest which may not be in the best interest of the overall system. In other words, when their self-interests are in conflict with the goals of the overall system, they may not subordinate their self-interest to achieving overall system goals.

Complex systems such as smart manufacturing and healthcare enterprises tend to be systems-of-systems (SoS), involving a huge number of independent stakeholders with varied interests, at each level in the organization hierarchy, with different specialties, and in different states. The traditional reductionist approach does not work for such complex engineered systems. Instead, a holistic approach is needed that specifically models system complexity. The complexity of a network or a SoS is a key metric used to determine the state of a system. Determining the complexity of a SoS network involves defining the state of the SoS network. The latter is concerned with identifying the nodes participating in specific transactions among nodes. Then, information theory can be used to calculate the totality of binary questions that need to be answered to determine the state of the SoS network. Given estimates of conditional probabilities of a node being involved in a transaction, complexity can be calculated and expressed in terms of bits. Rouse [5] presents an assessment of the complexity of the healthcare enterprise as comprising 27 bits based on five markets. This implies that determining which nodes (i.e., enterprises) are involved in any particular healthcare transaction is in the order of 1 billion binary questions – a huge undertaking to simply determine the state of the overall healthcare system. Comparing consumer complexity to total complexity in five sectors (i.e., aerospace, automotive, retail, healthcare, and telecommunications) is quite illuminating. Retail is by far the most complex; however, the consumer is saddled with a relatively small fraction of this complexity. In sharp contrast, the telecommunications industry has the worst ratio. Healthcare falls somewhere in the middle. Automotive has an impressive ratio just behind healthcare. And aerospace is slightly worse than automotive.

In light of the foregoing, complex engineered systems need to be designed to the extent practicable to suppress undesirable emergence. To achieve this objective requires including all stakeholder organizations (i.e., customers, partners, collaborators, channels, competitors, and regulators) in the enterprise model. Then, the overarching objective function becomes that of increasing complexity where it can be managed best while decreasing complexity for end users/consumers.

It is important to note that since it is not feasible to control the state of complex engineered systems, it should not be assumed that the governing enterprises would be able to manage the complexity of complex engineered systems. Therefore, the focus should be on managing complexity. Specifically, the design should make provision for monitoring and influencing system state, performance, and stakeholders. Complexity management can be facilitated by introducing sufficient flexibility for rapid redeployment of resources to exploit opportunities and resolve issues.

Incorporating flexibility in system design and program schedule is key to cost-effectively coping with uncertainty and exploiting opportunities. However, flexibility like all nonfunctional requirements comes at a price. Thus, a quantitative approach to valuing flexibility is needed. Real options, the engineering equivalent of financial options, can provide this capability. Incorporating real options at key points in the schedule and design facilitates exploitation of suddenly appearing opportunities while serving as a hedge against uncertainty.

Complex systems can be engineered only to a degree, because system characteristics such as learning, adaptation, and self-organization pose new challenges to traditional design. Their management also poses a challenge, because complex systems continue to redesign themselves through self-organization. As such it is important to recognize the differences between traditional systems and complex engineered systems. Traditional systems are managed to minimize costs. However, complex systems are managed to facilitate adaptation and learning, sustain resilience, and maximize value. Value pertains to the benefits afforded by the outcomes, not the outcomes themselves. Value also implies relevant, usable, and useful outcomes.

Therefore, it is imperative that stakeholders understand and appreciate the concept of complex system management and its implications.

There are several other considerations that also need to be addressed. First, there is no central authority in charge of a complex engineered system. This means the behavior of the system can only be influenced/shaped (through, e.g., incentives) rather than controlled (through power). Furthermore, since stakeholders are not employees but individuals with vested interest, traditional command and control need to be replaced with incentives and inhibitors/disincentives. In other words, stakeholders exhibit desired/appropriate behaviors in response to incentives, rather than comply with edicts and dictates.

Second, since the behavior/activities of stakeholders cannot be directly observed, their activities are not amenable to being managed. In this case, the value of their outcomes can be used as proxy. Unlike for traditional systems, where the focus tends to be on efficiency, the learning ability and adaptability of a complex engineered system can enable resilience and agility, rather than resorting to optimization based on outdated requirements.

Third, contractual relationships exist in complex systems, especially system-of-systems (SoS) in which the contributing systems have different governances. In such loosely coupled collaboration, stakeholders can change their allegiances at the end of their temporary contracts. In this regard, mutual commitments can dramatically lower the risk of such eventualities. For mutual commitments to endure, the agents (e.g., systems in a SoS) need to have close and trusted relationships, in conjunction with transparent organizational policies, practices, and outcomes.

Fourth, it is important to realize that while work is done by heterarchies, permissions and resources are provided by hierarchies. To the degree that the heterarchy has to pause operation to request permission and resources from the hierarchy, system efficiency and effectiveness are undermined. Thus, decision authority and resources should be delegated to the heterarchy with the right incentives and disincentives to evoke desired behaviors.

Fifth, organizational design doesn't really apply to complex engineered systems in that the organization is influenced and shaped by stakeholders as they "learn and adapt" to changing conditions. So, even if an initial organization was put into place, that organization would cease to exist after a series of adaptations. As such, management needs to maintain awareness of the organizational changes and manage to the prevailing organization to derive maximize value. In this regard, a management dashboard can provide a useful service [6].

2.2.8 Thinking in Stories

Many stakeholders are non-engineers and therefore unfamiliar with engineering modeling notion. These individuals tend to get left out in up-front engineering when system requirements are discussed and concept of operations is explored. As important, not everyone can relate to information presented as facts and numbers.

Table 2.5 Questions in constructing system stories

• Who interacts with the system?
• What do the interactions produce?
• When (i.e., under what conditions) do the interactions occur?
• When could the interactions happen again?
• Where does the interaction occur?
• Where else (in space and time) could the interactions occur again?
• Why does the interaction occur?
• What is the impact of such interaction on system behavior?
• What is the impact of such interaction on business or mission (story) outcome?

Specifically, facts and figures are not effective in explaining a product or process, because they don't reflect context. This is where storytelling comes. Thinking in terms of stories contextualizes facts and figures while facilitating collaboration among stakeholders in up-front engineering. According to Jonah Sachs, author of *Winning the Story Wars*, "A story takes all the senseless data that the world provides and turns it into something meaningful." Stories also foster abstract thought which is key to innovation, problem formulation, and problem framing. Thinking about systems in terms of stories is a structured process that consists of answering several key questions related to interactions with the system (Table 2.5).

Stories can be delivered as a lecture or interactively. Stories can be conveyed in the classroom by an instructor or interactively by a computer program. Stories can be delivered interactively in visual format or in transmedia environments. Visual storytelling is an inexpensive and effective means to communicate concepts, facts, and figures and foster collaboration among technical and nontechnical stakeholders. In visual storytelling, basic facts (e.g., caption, context, system) are included as metadata. Also, any reasonable assumptions made by stakeholders are accepted as true. Multiple views are typically employed to enhance collaboration. Stakeholders can view and interact with stories using role-specific lenses and function-specific lenses. Such lenses and filters help reduce system complexity by exploiting the "separation of concerns" principle. Visualization goes hand in hand with storytelling and interactive story execution. Specifically, visualization can be exploited to show new interactions or familiar interactions in unfamiliar but intuitively appealing ways. It is important to have a skeletal system story in mind to inform partial scripting of the story to impose a modicum of structure when stakeholders begin to explore alternate futures and system behaviors under different assumptions and conditions. Systems engineers have to go beyond thinking in terms of merely modeling the system using a traditional system modeling language. They need to think in terms of human-system interactions, embedded cues and prompts, alternate change propagation paths, and story injects. They need to bring this thinking to bear in constructing purposeful stories for execution in virtual worlds. System stories ensure that all stakeholders understand system behaviors from their respective perspectives (with different assumptions and conditions). Stories alleviate uncertainty and doubt. They help users cope with change, which can be unsettling. They help users appreciate and understand the need for compromise and how to frame and pursue trade-offs. Stories help simplify complex systems and problems. They can persuade when facts alone are not enough. They help stakeholders reach desired conclusions by evoking strong imagery. In sum, thinking in terms of stories eliminates extraneous constraints, engenders strong imagery, fosters creative option generation, and reduces uncertainty and doubt through exploration of alternative futures and shaping desired outcomes.

2.2.9 *Thinking About Humans and Technology*

Technological advances, operational demands, and system complexity are redefining the roles of humans in relation to systems. Specifically, system adaptability requirements, automation, and social media are redefining the role of humans in relation to systems. System adaptability and SoS reconfigurability requirements impose additional cognitive load because of the need to context switch and multitask. While automation continues to elevate the role of humans from primary task performers to supervisors (with over-ride privileges), the unintended consequence is an erosion of skills needed to backup automation in case of malfunction or failure. And finally social media and crowdsourcing have literally changed the very definition of collaboration and collaborators. It is against this backdrop that we will explore the changing roles of humans and the challenges of human-systems integration especially for adaptable systems.

The potential for "disconnect" between people and technology is well documented in the literature for consumer products and large-scale systems literature [2, 7, 8]. The Patriot missiles, deployed in the 2003 Iraq war, offer a compelling illustration of this disconnect. Operators of this missile were trained to trust the system's software because the Patriot missile is a highly automated system. Such trust is essential especially when operating in a heavy missile attack environment [3]. It is important to recognize that this was not the case in the Iraqi battlespace in which the missile batteries were operating in an environment sparsely populated with missiles but with several friendly aircraft. The missile operators, inadequately trained, were unaware that the Patriot radar system was susceptible to recording spurious hits and occasionally issuing false alarms (i.e., mistaking friendly aircraft for enemy missiles) without displaying the uncertainty in target identification. Not surprisingly, these operators trusted the system's assessments and missile launch decisions against potentially hostile targets. These factors contributed to the unfortunate shootdown of a British Tornado and a US Navy F/A-18. A Defense Sciences Board study concluded that "more operator involvement and control in the function of a Patriot battery" was necessary to overcome the system's limitations [3]. Despite this recognition, system operators continue to be unfairly blamed for systemic failures.

This fact did not go unnoticed by Chiles [2] who cautioned, "Too often operators and crews take the blame after a major failure, when in fact the most serious errors took place long before and were the fault of designers or managers whose system would need superhuman performance from mere mortals when things went wrong."

The primary design flaws that Chiles refers to were largely failures in proper coordination of interactions between people and technology during system development and operation [9]. In recent years, the need for systems to become increasingly more adaptive to cope with changes in the operational environment has made the integration of humans with software and systems even more challenging. In response to these challenges, the DoD made a concerted push to incorporate human considerations into the systems engineering life cycle [10]. This emphasis, in large part, led to the creation of the new multidisciplinary field of human-system integration (HSI) [11, 12]. HSI is the study of interactions between humans and systems to produce human-system designs that are compatible, safe, consistent, and efficient. These interactions continue to become increasingly more complicated as human roles continue to evolve from that of an operator outside the system to that of an agent within the system. Compounding the problem is the fact that misconceptions about what it takes to integrate humans with software and systems continue to linger in the software and systems engineering communities [13]. Perhaps the single biggest misconception is that humans are "suboptimal job performers." This mindset leads to software and systems that are specifically designed to shore up or compensate for human shortcomings. With this mindset, it is hardly surprising that humans are forced to operate or work within systems that are inherently incompatible with their conceptualization of work.

Humans have specific strengths and limitations that need to be well understood before determining how best to integrate them with software and systems [14–17]. The key findings from the literature that bear on human-system integration are presented in Table 2.6.

The literature on human-machine systems offers ample evidence that poorly designed automation can produce performance degradation of the overall human-machine system. An important aspect of such performance degradation is the lack of "fit" between the mental models of humans, cognitive demand of the work environment, and automation design. Poorly designed automation can degrade human performance (Table 2.7).

2.2.10 (Re)Thinking Assumptions

All too frequently, system designs are based on unstated and occasionally unwarranted assumptions about human behavior. These assumptions can often lead to unintended consequences and give rise to systemic failures. The following examples of unexpected outcomes and unintended consequences can be traced directly to unwarranted assumptions about human behavior.

Table 2.6 Key HSI findings

• **Human performance** [18–21] – Varies nonlinearly with several factors – Follows an inverted U-curve relative to stress and vigilance – Excessive cognitive complexity can lead to poor performance on secondary task and eventually task shedding [18]
• **Human error** [18, 20] – Lack of inspectability into system operation can induce human error – Incompatibility between human processes and machine algorithms can lead to human error – Sustained cognitive overload can lead to performance degradation, fatigue, and human error
• **Human adaptivity** [22, 23] – Adaptivity is a unique human capability that is neither absolute nor perfect – Humans do adapt under certain conditions, but usually not quickly – Human adaptation rate sets an upper bound on how fast systems can adapt – Trade-off between human adaptation rate and error likelihood – Need to define acceptable error rate (context-dependent) to set upper bound on system adaptation rate
• **Multitasking** [22, 23] – Humans do not multitask well – High multitaskers have difficulty filtering out irrelevant information, can't compartmentalize to improve recall, and can't separate contexts (Stanford research)
• **Decision making under stress** [22, 23] – Humans simplify environment by disregarding/underweighting complicating factors – Reduced ability to process multiple cues or perform trade-offs
• **User acceptance** [18, 22, 23] – Overly complex system design can lead to rejection of the system – Understanding software/system operation is not necessary to develop trust in system
• **Risk perception and behavior** [2, 3] – Humans accept greater risks when in teams – Humans have a built-in target level of acceptable risk
• **Human-system integration** [2, 3] – Humans are rarely exactly right; however, human errors tend to be relatively minor – Software/system solutions tend to be precisely right, but when wrong they can be way off

In 2008, a Metrolink commuter train crashed headlong into a Union Pacific freight locomotive after going through four warning lights. The engineer (i.e., the driver) failed to hit the brakes before the train crashed. A teenage train enthusiast later claimed to have received a cell phone text message from the driver a minute before the collision [8]. So, was the Metrolink train accident a human error, a systemic problem that manifested itself as a human error, or both? The answer is *both*. Since the driver was doing a split shift, he was clearly tired. He was also multitasking. Humans don't multitask well and are error-prone in such circumstances. However, the system was also not designed for integration with the human in that

Table 2.7 Poor automation design can degrade human performance

• **Cognitive load in supervising automation** [24, 25] – Load in monitoring automated task performance can outweigh automation benefits
• **Automation-induced complacency** [26] – Over-reliance on automation can increase errors as humans begin to rely on automated cues rather than vigilant information seeking and cognitive processing [27]
• **Partially automated system with incomplete knowledge** [28] – A system, operating outside its competence regime, stays in the loop to critique operator performance based on erroneous assessment of work constraint violations
• **Mistrust of automation** [29] – Can lead to disuse, neglect, underutilization – Typically arises from poor design (e.g., high rate of false alarms in an alerting system)
• **Erosion of operator's expertise and engagement** [30] – Inappropriate automation can lead to skill decay or dysfunctional skills – Operator unable intervene effectively when automation malfunctions

the system design unrealistically assumed an optimal human, i.e., one who could multitask, one who would not tire, and was a goal-driven maximizer of personal utility. This was an accident waiting to happen [13]. In this case, unwarranted assumptions about the human led to the tragic accident [31]. For example, assuming that humans are optimal information processors can lead to dire results because humans do tire and don't multitask well.

Another instance of where erroneous assumptions can be made is in architectural and algorithm selection in a human-machine system. Specifically, the human role in relation to the system plays a significant role in system architecture design and algorithm selection. For example, if the human is expected to be replaced by automation in the future, then the system architecture would emphasize a different set of quality attributes than if the human role was intended to be integral to the system (i.e., permanent). The same is true of algorithm selection. Consider the selection of a route planning algorithm for an autonomous ground vehicle. Invariably, a constrained optimization algorithm would be used to solve the route planning problem. Now consider route planning for a human-supervised ground vehicle in which the human needs to specify waypoints along the way. In this case, the algorithm needs to be interactive, inspectable, and understandable so that the human can intervene to specify way points. As such, a heuristic algorithm becomes preferable to the optimization algorithm because the heuristic algorithm allows the human to understand system reasoning and intervene effectively [31]. In this example, algorithm inspectability is more important than algorithm optimality. In fact, this is a key trade-off.

Thus, the role of the human in the overall system is key to architectural paradigm and algorithm selection. Specifically, it is important to determine whether the human is central to system operation or merely an adjunct or enabler to be replaced by automation in the future. Fourth, system architects need to focus on combined human-system performance, not the performance of each in isolation. This also means that the focus should be on combined metrics, not individual metrics. And, finally, a change in the operational environment can potentially change how people perceive and compensate for risks [13].

2.2.11 Thinking Software Development

In *Augustine's law*, Norm Augustine offered a thought-provoking aphorism about software. Software is like entropy. It is difficult to grasp, weighs nothing, and obeys the second law of thermodynamics, i.e., it always increases [32]. In the twenty-first century, software will continue to "grow" in a sociotechnical ecosystem comprising customers, end users, developers, maintainers, testers, and other stakeholders. Their continued participation is crucial to software acceptance both in the government and the commercial sector. In the recent past, software has been a process-driven product. However, with increasing software complexity, it is becoming apparent that the people aspect of software deserves greater attention and emphasis. The people aspect comprises people decisions, personnel skillset, training, motivation, creativity, and talent.

Software has been a process-driven product for the last few decades. This view has inadvertently de-emphasized the importance of people in the software life cycle [7]. The reality today is that:

People with appropriate training perform software-related activities, often subject to governing standards and legacy constraints within development environments to achieve desired outcomes.

Today with ever-increasing software sophistication, human ingenuity is being challenged like never before. No longer does it suffice to just follow a disciplined development process because people are becoming increasingly crucial in performing trade-off analysis and in creating a satisfying user experience [2, 7]. In addition, people are key to ensuring that software performance, quality attributes, schedules, and cost objectives are being met. Exclusive focus on software process can potentially stifle human creativity and inhibit human contributions throughout the software life cycle. Furthermore, as software continues to grow in complexity and humans continue to become more and more an integral part of software-based systems, predictable software behavior is becoming crucial to software system safety [8].

Today the proportion of software in systems continues to increase dramatically. This recognition has led to the creation of the term "software-intensive systems." People contribute in a variety of ways to software-intensive systems. For example, humans create new paradigms, explore the software design trade space, discover patterns and trends, provide decision rationale, attempt to explain anomalous behavior, and assure smooth integration of people and software. Yet, the importance of people in the software life cycle continues to be underemphasized. This is surprising in that software is largely a people creation that is maintained, supported, and adapted by people. People are also responsible for software quality, and yet scant attention is devoted to the talent, training, creativity, and motivation of people

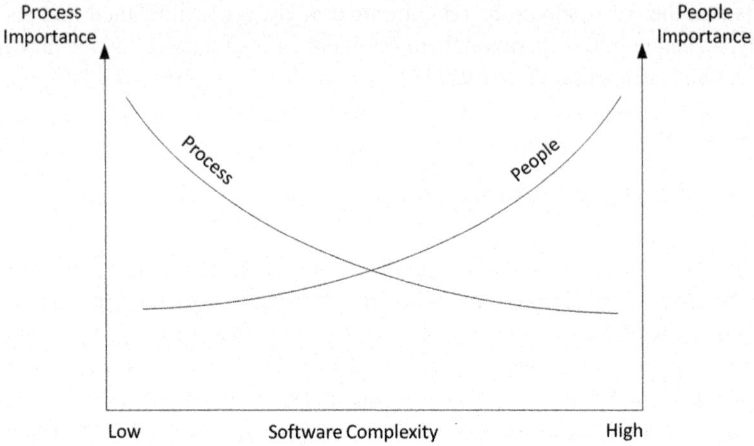

Fig. 2.1 Increasing software complexity driving paradigm shift

responsible for assuring software quality [33]. Clearly, process will always play an important role but more as an enabler than a driver.

There are several compelling reasons to make people the primary focus in software development today. First, software is a creation of people and quite frequently for the use of people. Exclusive focus on process can stifle creativity and compromise user acceptance. Second, safety is becoming an increasingly important consideration in software-intensive systems. Safety subsumes predictable software behavior in the face of disruptive events [8]. It is important to note that processes do not automatically address safety concerns. It is people who introduce safety concerns in the software life cycle. Third, with the need for adaptive processes (e.g., agile) and the need for adaptable systems (to survive and operate in changing operational environments), the shift toward people-driven development is becoming inevitable [10–12]. Finally, with the advent of multi-domain software that cuts across multiple domains (e.g., electrical, optical, mechanical) and multiple disciplines (e.g., physics, social sciences, cognitive science), software complexity has increased dramatically. Collectively, these trends speak to the need for people-driven, process-enabled software development and use.

Figure 2.1 presents a notional graph illustrating the approximate relationships between process importance and software complexity and between people importance and software complexity. As shown in this figure, as software complexity increases, software development becomes less and less process driven and more and more people driven, albeit process enabled. A key implication of this trend is that if the developing organization expects software to grow in scale and complexity, the organization is better off adopting people-driven, process-enabled software development practices [3, 7, 8, 13].

In the recent past, several developments have collectively pointed to a much-needed shift from process-driven to people-driven software development. First and foremost is the uncertainty about the operational environment, rate of maturation of promising technologies, and personnel turbulence resulting from retirements, layoffs, and personnel moves. Second, software is becoming more complex because of ever-increasing scale and ever-growing need for adaptability in light of the changing roles of humans in relation to software. These trends are being driven by the need for systems to be long lived and capable of coping with unknown operational environments. Third, organizations are turning to adaptive processes such as agile development paradigm, which is being viewed as a source of competitive advantage when applied correctly. It requires an accomplished team of developers, effective leadership in pulling the team together, and a change in mindset associated with traditional process-driven development in which roles are important but individual people are viewed as interchangeable/substitutable parts, with people availability trumping people skillset [9, 13, 14].

Alistair Cockburn, in his book *Characterizing People as Nonlinear, First Order Components in Software Development*, argues that *predictable processes* require components with *predictable behavior*. And, people are anything but predictable. Treating humans as interchangeable components or replaceable parts in software development is a misjudgment. Human behavior tends to be variable and nonlinear. Humans exhibit an uncanny ability to succeed in novel ways while also exhibiting a disconcerting capacity to fail in unimagined ways. It is the failure to account for these factors in software development that inevitably result in schedule and cost over-runs. In fact, it is fair to say that humans strongly figure in both project successes and failures [8].

Unfortunately, the mistaken belief that people are interchangeable resources is deeply ingrained in business thinking. It dates back to Frederick Taylor's scientific management approach for performing repetitive tasks such as running a factory [16]. However, for highly creative work such as software development, this view is clearly inapplicable. And today, with the advent of smart manufacturing, manufacturing also no longer abides by this tenet. Another key tenet of Taylor's theory is that the people doing the work are not best suited to determine how best to do the work. While this tenet may hold, to a degree, on the factory floor, it is untrue of software development. In fact, people attracted to software engineering tend to be the best and the brightest, with the culture of youth pervading the field [8, 16].

So, what is it that people bring to software? People bring imagination, novel insights, storytelling ability, and an uncanny ability to discern and exploit patterns [2, 3]. These capabilities have the potential to transform software development in unprecedented ways to achieve dramatic improvement in software quality, responsiveness, cycle times, and life cycle costs.

A people-driven, process-enabled view of software goes well beyond the process perspective. It is sensitive to business concerns and constraints, implications of software-related decisions on short-term, mid-term, and long-term concerns of a program or business. It is cognizant of the available skillset in both management and development teams. It shows understanding of programmatic and technical trade-offs and the importance of collaboration and full stakeholder participation in the software life cycle. The latter is essential for reasoned compromise that addresses stakeholders' concerns and resolves issues. It is also essential for stakeholder acceptance of collaboratively made decisions and elimination of extraneous design iterations and rework [7]. The maturity and experience of team members and leadership play a pivotal role in defining use cases, specifying architecture, and developing the right set of abstractions.

From Process-Driven to People-Driven, Process-Enabled Development People-driven development is more than stakeholders influencing and agreeing on what is being created. It is more than empowering engineering teams and the activities they perform to develop software. And it is more than directing software users in the use of software. It is in fact all of the above. People-driven development means humans playing an active role in software-related trade-offs, designing the software, managing the software development process, and even distributing software development activities to the development team members. People-driven development is also influenced by culture and power distance [17]. Compounding the problem is the "clash of values" between developers and program managers [18]. And, of course, human behavior exhibits nonlinearity and variability [2, 19]. These factors influence both the development process and the software product. Cockburn [19] and Madni [2, 3] identify specific factors that influence the outcome: humans are social beings who perform best in face-to-face collaboration; humans are inconsistent and inconsistency shows up over time; humans exhibit variability from day to day and place to place; and humans generally want to do the right thing for their organizations.

These characteristics bear directly on process. It is important to recognize that process enforcement can vary from strict to loose. In light of human characteristics and ever-growing system complexity, loose process enforcement is preferable to strict enforcement. In cases, where strict process enforcement is required, there is a need for performance support for humans to behave consistently.

Software life cycle processes provide a structured disciplined means to guide the development of complex, real-world software [20]. This software spans: primary processes (acquisition, supply, development, operation, maintenance), supporting processes (documentation, configuration management, quality assurance, reviews and audits, problem resolution), and organizational processes (management, infrastructure, maintenance, improvement, training). The question that needs to be asked in where do life cycle processes benefit software design and where do they become an impediment. For most support and organizational processes, following software life cycle process is a benefit. Also, periodic architecture and design reviews help to ensure design quality and traceability between requirements and design elements to ensure design completeness. However, there are times where strict process enforcement becomes a hindrance to creativity and innovation [21]. In this case, humans can "dial back" on strict process enforcement and adopt loose process enforcement. This shift puts people in charge of the process, making it people-driven, process-enabled software. This recognition is at the heart of adaptive software development in general and agile development, in particular.

Agile processes (or agile, for short) are a prime example of people-driven, adaptive development. Agile relies on process acceptance by the development team, not process imposition by management [10–12, 15]. In other words, only developers themselves can choose to follow an adaptive process. This is especially true of extreme programming (XP), which requires disciplined execution, with developers making all the decisions and generating all time estimates. This is a huge cultural shift for management in that it requires sharing of responsibility between developers and management [15].

2.2.12 Thinking Platforms

The purpose of the platform is to accomplish effective matches among users and facilitate the exchange of goods, services, and social capital, thereby enabling "win-win" for all participants. Platforms are networked resources capable of delivering solutions, whose value increases in proportion to the number of users of that solution. Thus, a smart phone with a single user has little value. However, as other users enter into the picture, and as more apps are added, and as social networks comprising the smart phone grow, the value of the smart phone increases dramatically. The key point here is that the real value stems from "network effects" beyond that provided by the smart phone. In their landmark book, *Platform Revolution*, Parker, Van Alstyne, and Choudhary make a compelling case for how hyper-connectivity is transforming the economy and how network effects can be leveraged to create and sustain competitive advantages for organizations.

Platform strategy, a shift from product strategy, is intended to exploit hyper-connectivity and resultant networks that define and underlie social and economic advances in the twenty-first century. The shift from a *product strategy* to *platform strategy* has its roots in business strategy for the twenty-first century. It is all too apparent today that product companies such as Blackberry and Nokia have been unable to compete with platform companies such as Google, Apple, and Amazon. It is also important to note that the platform strategy is not limited to businesses that are merely websites or mobile applications. The platform strategy is a general strategy that is being pursued today by companies that are websites, sell mobile applications, or sell physical products. So, what is a platform strategy?

A platform strategy exploits hyper-connectivity and network effects to deliver value to users. Specifically, it exploits technology to connect people, organizations, and resources within a dynamic ecosystem in which value creation emphasis is outwardly focused, i.e., external value creation. This is in sharp contrast to value creation in product companies that tends to be inwardly focused. Unlike a traditional business that owns its resources and employs a linear value chain spanning producers to consumers, a platform business harnesses the resources of others and is able to scale at rates that cannot be matched by a traditional business. For example, Uber does not own cars, and Airbnb owns no real estate. And yet, each is growing at a rate that cannot be matched by their traditional counterparts.

According to Parker et al. [34], the key assets of platform businesses are *information* and *interactions* that collectively constitute the source of value creation and competitive advantage. Both traditional businesses focused on reinventing themselves and startups looking to ramp up quickly stand to benefit from a platform strategy. Of course, the transition to a platform business takes more than just a technical and business transformation. Issues of access, governance, pricing, liability and other related factors have to be addressed as well. And they are being addressed by businesses which have successfully made this transition, or are in the process of doing so.

Platforms, and specifically digital platforms, can be classified as transaction platforms and innovation platforms. While both exploit hyper-connectivity and the power of the Internet, they employ different value creation strategies. *Transaction platform* companies create value by facilitating transactions between different types of individuals and organizations that would otherwise find it difficult to locate each other. *Innovation platform* companies generate value by co-innovating and cocreating products and services with other firms that belong to their platform ecosystem.

Examples of transaction platforms are Facebook, Google, Amazon, and Apple. Examples of innovation platforms are Microsoft, Salesforce, SAP, and Intel. Innovation platform companies such as Oracle and SAP, that have historically relied on system implementation and consulting partners, are beginning to reduce their reliance on software implementation partners as they embark on developing private cloud solutions that can be implemented quickly and upgraded easily while delivering a better return on investment [35]. To improve platform value, these companies are changing their curation policies. In Platform Revolution [34], *curation* is defined as the process by which a platform filters, controls, and limits user access to the platform, the activities they participate in, and the connections they form with other users. Ultimately, platform companies offer several benefits to the economies they serve. First, they increase productivity through efficient "matchmaking" between consumers and providers and more efficient asset utilization. Second, they are a key source of *innovation* and *job creation*. Third, they embrace *decentralized control* over central control with significant value generation occurring outside traditional organization boundaries, with supply chain managers effectively becoming curators of network value. In sum, a platform business generates value through interactions with external producers and consumers. The platform provides an open, participative infrastructure for these interactions and sets governance conditions for them.

2.3 Summary

The demands of the twenty-first century require a new mindset to make sense of the world around us, exploit opportunities, and manage complexities. In particular, ten new perspectives on how we need to think to survive and thrive in a complex, hyperconnected world are presented:

- Thinking about thinking – think in terms of the big picture, interconnectedness, asking "smart questions," and identifying leverage points

- Thinking about humans – not as suboptimal job performers that need to be shored up or compensated for but as sources of creativity that need to be exploited; the key is to exploit human capability to adapt without exceeding human capacity to adapt [2]
- Thinking about models – as organizing constructs and as means to conquer complexity, acquire insights, understand patterns and trends, and focus data collection for analytics
- Thinking trade-offs – to balance competing stakeholder demands and conflicting systems objectives subject to a variety of constraints
- Thinking change propagation – within a complex system and its impact on system behavior – change is absorbed, change produces oscillations, or change amplifies as it propagates often leading to undesired system behaviors and unintended consequences
- Thinking wicked – some problems are resistant to solutions in the traditional sense; need to recognize that some problems can change with interventions intended to solve them; need to understand that such problems should to be handled differently
- Thinking complex systems – this means thinking in terms of holism, self-organization, agility and resilience, scaling up methods, influence, incentives, adaptation and learning, and mutual commitments
- Thinking in terms of stories – exploiting context provided by stories to understand and simplify system, communicate perspectives, and reach reasoned compromise and consensus
- Thinking humans and technology – understand human relationships with technology, need for mutual adaptation and learning, need for mutual commitments not contracts, and need to understand strengths and limitations of each
- Thinking software development – shift from process-driven development to people-driven, process-enabled development
- Thinking platforms – need to think in terms of networked resources capable of delivering solutions; solution value is proportional to the number of uses and users of that solution; both information and interactions are sources of value creation

This foregoing shift in thinking is key to coping with hyper-connectivity and complexity and essential for exploiting convergence.

References

1. Neumann JV. Mathematical foundations of quantum mechanics. Princeton: Princeton university press; 1955.
2. Madni A. Integrating humans with and within software and systems: challenges and opportunities. Invited paper CrossTalk. J Def Softw Eng. 2011:4–7.
3. Madni AM. Integrating humans with software and systems: technical challenges and a research agenda. Syst Eng. 2010;13(3):232–45.

4. Box GE, Draper NR. Empirical model-building and response surfaces. New York: Wiley; 1987.
5. Bodner DA, Rouse WB. Understanding R&D value creation with organizational simulation. Syst Eng. 2007;10(1):64–82.
6. Madni AM, Richey M, Ordoukhanian E, Venkatesh J, Zender F, Chang K, et al. Exploiting storytelling in collaborative systems engineering: towards a smart experiential dashboard. Alabama: Conference on Systems Engineering Research Hunstville; 2016.
7. Madni AM. Expanding stakeholder participation in upfront system engineering through storytelling in virtual worlds. Syst Eng. 2014;18(1):16–27.
8. Cockburn A. People and methodologies in software development: University of Oslo Norway. 2003.
9. Madni AM. Thriving on change through process support: the evolution of the ProcessEdge Enterprise suite and TeamEdge. Inf-Knowl-Syst Manage. 2000;2(1):7–32.
10. Cockburn A. Agile Software development: the cooperative game:Pearson Education. 2006.
11. Madni AM. AgileTecting™: a principled approach to introducing agility in systems engineering and product development enterprises. J Integr Des Process Sci. 2008;12(4):49–55.
12. Madni A, Agile systems architecting: Placing agility where it counts. Conference on Systems Engineering Research (CSER). 2008.
13. Madni AM, Spraragen M, Madni CC. Exploring and assessing complex systems' behavior through model-driven storytelling. 2014 IEEE International Conference on Systems, Man, and Cybernetics (SMC); 2014/10: Institute of Electrical & Electronics Engineers (IEEE). 2014.
14. Austin RD. Measuring and managing performance in organizations. New York: Dorset House Publishing Co., Inc.; 2013.
15. Taylor FW. Shop management. New York and London: Harper & Brothers Publishers; 1911.
16. Taylor FW. The principles of scientific management. New York: Harper; 1914.
17. Hofstede G, Hofstede GJ, Minkov M. Cultures and organizations: software of the mind: Citeseer. New York: McGraw-Hill; 1991.
18. Cockburn A, Software development as Community Poetry Writing. Cognitive, cultural, sociological, human aspects of software development. Annual meeting of the Central Ohio chapter of the ACM; 1997.
19. Cockburn A, Characterizing people as non-linear, first-order components in software development. International Conference on Software Engineering 2000; 1999.
20. Suryanarayana G, Sharma T, Samarthyam G. Software Process versus Design Quality: Tug of War? IEEE Softw. 2015;32(4):7–11.
21. Madni CC, Madni AM. Web-enabled collaborative design process management: application to multichip module design. SMC'98 Conference Proceedings 1998 IEEE International Conference on Systems, Man, and Cybernetics (Cat No98CH36218): Institute of Electrical & Electronics Engineers (IEEE); 1998.
22. Adolph S, Cockburn A, Bramble P. Patterns for effective use cases. Boston: Addison-Wesley Longman Publishing Co., Inc; 2002.
23. Madni AM. People-driven Process-enabled software development: A 21st Century Imperative. CrossTalk: The Journal of Defense Software Engineering. 2016 (Cyber Workfore Issues).
24. Kirlik A. Modeling strategic behavior in human-automation interaction: why an "aid" can (and should) go unused. Human Factors:J Hum Factors Ergon Soc. 1993;35(2):221–42.
25. Sheridan TB. Telerobotics, automation, and human supervisory control: MIT press; 1992.
26. Parasuraman R, Molloy R, Singh IL. Performance consequences of automation-induced 'Complacency'. Int J Aviat Psychol. 1993;3(1):1–23.
27. Mosier KL, Skitka LJ. Human decision makers and automated decision aids: made for each other. In:Automation and human performance: theory and applications. New Jersey: Lawrence Erlbaum Associates Inc; 1996. p. 201–20.
28. Guerlain SA, Factors influencing the cooperative problem-solving of people and computers Proceedings of the Human Factors and Ergonomics Society Annual Meeting. SAGE Publications; 1993.

29. Parasuraman R, Riley V. Humans and automation: use, misuse, disuse, abuse. Hum Factors. 1997;39(2):230–53.
30. Klein G. Implications of the Naturalistic Decision making framework for Information Dominance. DTIC Document. 1997.
31. Madni AM, Sage AP, Madni CC. Infusion of Cognitive Engineering into Systems Engineering Processes and Practices. 2005 IEEE International Conference on Systems, Man and Cybernetics: Institute of Electrical & Electronics Engineers (IEEE); 2005.
32. Augustine NR. Augustine's laws: AIAA; 1997.
33. Madni A, Moini A, Madni C. Cross-Cultural decision making training using behavioral Game-Theoretic Framework. Advances in Cross-Cultural decision making: Informa UK Limited; 2010. p. 385–394.
34. Parker G, Van Alstyne M, Choudary SP. Platform Revolution: how networked markets are transforming the economy and how to make them work for you. New York: WW Norton company, Inc; 2016.
35. Evans PC, Gawer A. The rise of the platform enterprise: a global survey. In: The emerging platform economy series, The Center for Global Enterprise. New York: Springer; 2016.
36. Madni AM, Spraragen M, Madni CC. Exploring and assessing complex system behavior through model-driven storytelling. IEEE Systems, Man and Cybernetics International Conference, invited special session "Frontiers of Model Based Systems Engineering", San Diego, Oct 5–8, 2014.
37. Madni AM. Integrating humans with software and systems: technical challenges and a research agenda. 22nd Annual Systems and Software Technology Conference, Salt Lake City, April 27, 2010.

Chapter 3
Disciplinary Convergence

"The world is being re-shaped by the convergence of social, mobile, cloud, big data, communication and other powerful forces. The combination of these technologies unlocks an incredible opportunity to connect everything together in a new way and is dramatically transforming the way we live and work."

– Marc Benioff, internet entrepreneur

3.1 Typology of Convergence

A major consequence of the world becoming increasingly hyper-connected is that problems are becoming much too complex to solve using methods from a single discipline. This recognition in part is motivating the pursuit of convergence research. *Convergence* is an approach to problem solving that cuts across disciplinary boundaries. It often surfaces opportunities to achieve disruptive advances that can benefit society and human quality of life. A 2014 National Academies report [1] defines convergence as "an approach to problem solving that cuts across disciplinary boundaries." The report states that "convergence integrates knowledge, tools, and ways of thinking from life and health sciences, physical, mathematical, and computational sciences, engineering disciplines, and beyond to form a comprehensive synthetic framework for tackling *scientific* and *societal challenges* that exist at the *interfaces of multiple fields*." The report also offers multiple definitions and types of convergence that can potentially transform research and education in the 21st century.

The term "convergence" encompasses both technology convergence and disciplinary convergence. Technology convergence has been underway for more than a decade and has already paid rich dividends. In terms of how we communicate, learn, and socialize. The integration of telephony and data technologies is an early example. This integration consisted of placing voice network (telephony), video network (television, satellite), and the Internet (rich media) onto a common platform. Shortly thereafter, telephone, print media, film, radio, TV, and electronic (computer) data that used to be separate technologies that in time came together. Over the last two decades, technology convergence has produce increasingly more advanced and newer generations of multimedia communication products and platforms.

A.M. Madni, *Transdisciplinary Systems Engineering*,
DOI 10.1007/978-3-319-62184-5_3

Technology convergence has transformed entire industries such as the geospatial products industry. Not too long ago, remote sensing, AM/FM radios, computer-aided design, Geographic Information Systems, and Global Positioning system (for navigation) were distinct vertical markets served by vertical technology providers. Today, these technologies are converging rapidly with the resultant market becoming increasingly more horizontal, and stove-piped vertical solutions rapidly disappearing from the marketplace. Technology convergence is an enabler of disciplinary convergence.

With disciplinary convergence, the emphasis is on enhancing compatibility, exploiting synergy, and reconciling differences among different disciplines. Of necessity, the pursuit of convergence is evolutionary and nonuniform [2].

As convergence among disciplines intensifies, the contributing disciplines tend to enrich each other even as knowledge gaps between them begin to appear. Occasionally, new disciplines emerge. What is lacking today is a new way of thinking that strives to harmonize traditional disciplines by reaching beyond their traditional boundaries to identify and fill knowledge gaps and overcome incompatibilities [3]. This is what transdisciplinary thinking is about.

3.2 The Promise of Disciplinary Convergence

Disciplinary convergence is the unification of insights from distinct disciplines to solve problems more effectively than before. For example, the convergence of distributed collaboration, social media, and crowdsourcing is an example of disciplinary convergence. It changed the very definition of collaboration and collaborators. Social media exploits social science and the Internet. Crowdsourcing which is about soliciting and acquiring services, ideas, or content from a large number of people exploits ideation, social media, and the Internet. Historically, a long and tedious process, crowdsourcing has taken center stage with the advent of the Internet and the web. In fact, the term "crowdsourcing" gained popularity in the Internet era, despite being around for a long time. In the last few years, social media has become an enabler of crowdsourcing because it allows organizations to reach a wide audience rapidly to locate potential information providers.

Today, crowdsourcing and social media have become complementary and synergistic. Both owe their growing popularity to the Internet. Crowdsourcing appeals to our sense of creativity, while social media appeals to our need to interact and socialize. The combination of the two provides an effective platform for problem solving, question answering, and staying informed. More recently, crowdsourcing has become a platform for starting ventures, because it fosters out-of-the-box thinking by democratizing activities such as product development and startup launch. Social media empowers consumers/users to share needs and locate providers to satisfy them. It empowers consumers to directly link to product/service/information providers and other resources without incurring cost. It helps spread the word about crowdsourced projects and facilitates reaching potential contributors.

Table 3.1 Facets of convergence

• An approach to complex problem solving and answering questions – cuts across disciplinary boundaries
• The emerging of ideas, approaches, and technologies – from widely diverse fields of knowledge – at a high level of integration
• Formation of a web of partnerships – support convergent scientific investigations – translate resultant advances into new products and innovations
• A cultural shift from discipline-based departments to theme-based organizations – open, inclusive culture – practitioners conversant across disciplines
• Merging of expertise (underway since the 1920s)
• Common sets of concepts, metrics, and understanding of goals

At its core, disciplinary convergence is concerned with the integration of knowledge, tools, and thinking approaches from physical and life sciences, mathematics, computer science, and engineering fields to tackle scientific and societal problems. In the scientific community, convergence also refers to the formation of a web of partnerships needed to support convergent investigations and translate the resultant advances into new products and innovations [1]. From an organizational perspective, convergence is a cultural shift from discipline-based to theme-driven organizational units. Of course, the latter comes with a new set of challenges when it comes to administration, research, teaching, financial infrastructures, and partnership agreements and mechanisms. The different facets of convergence are presented in Table 3.1.

Disciplinary convergence began to gain momentum in the last decade. It began with our increased understanding of the molecular basis of life and the twenty-first century being dubbed the "century of biology." Both genomics and bioinformatics were largely a result of unraveling the structure of DNA and the ability to sequence, compare, and interact with the human genome at modest costs [4]. The integration of life sciences with other disciplines is likely to continue well into the twenty-first century. The central idea of disciplinary convergence is that bringing together concepts, approaches, technologies, and thinking approaches from diverse disciplines helps in the formulation and solution of complex problems. For example, genomics resulted from a combination of genetics, chemistry, optics, and bioinformatics [4]. Ultimately, the pursuit of disciplinary convergence requires an open mind, an inclusive culture, and willing participants.

The foregoing discussion points to increasingly higher levels of convergence with technology-enabled disciplinary convergence being at the higher end of the convergence spectrum. These advances are the key enablers of transdisciplinary systems engineering as the systems engineering research community collectively

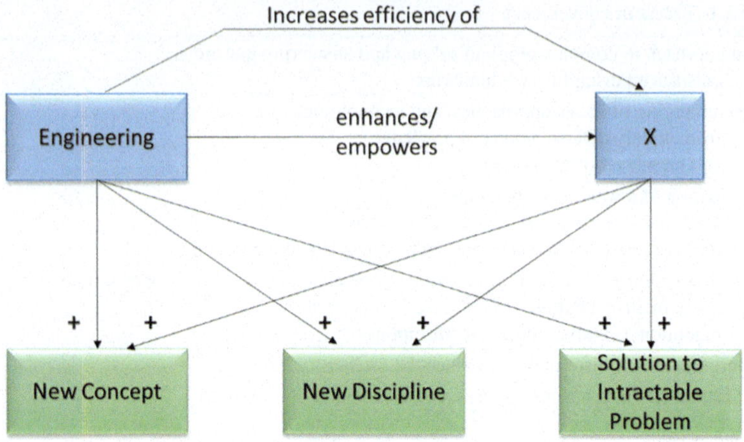

Fig. 3.1 Types of disciplinary convergence (engineering and X)

reaches beyond engineering to other disciplines to find connections [3]. In this regard, the interaction of engineering with another discipline (X) can produce several desirable outcomes (Fig. 3.1).

Engineering Makes X More Efficient In this case, X is collaboration among team members (technical and nontechnical). Engineering can introduce visualization, interactive storytelling, and experiential interfaces to make collaboration more effective and efficient.

Engineering Empowers or Enhances X X, for example, is traditional collaboration (face-to-face and distributed). Social media, a product of engineering, enhances traditional collaboration by expanding the pool of collaborators to include strangers ("crowdsourcing") capable of contributing to the discussion while also expanding the definition of collaboration to include both known individuals and "strangers."

Engineering and X Produce a New Discipline This is the case with engineering and biology. The unification of these disciplines produced the discipline of bioengineering. Similarly, the coming together of computer science and sociology introduced the discipline of computational sociology.

Engineering and X Produce a New Concept In this case, engineering and the concept of financial options (X) from economics produced a new concept called real options. In the same vein, engineering and the human immune system (biology) gave rise to the concept of "artificial immune system." The latter provided a convenient metaphor and analogy for architecting cyber-secure, resilient systems.

Engineering and X Together Solve an Intractable Problem Quite often problems viewed exclusively through an engineering perspective may appear intractable. However, upon combining engineering and another discipline, X, the problem can become tractable. Problems such as climate change, healthcare enterprises, and adaptive energy grids fall into this class of problems.

Disciplinary convergence is already beginning to make its mark in sociotechnical system design, engineering, and education. Disruptive collaboration, experiential design, immersive technical storytelling, and cross-disciplinary methods for stimulating novel option generation are all examples of disciplinary convergence. *Disruptive collaboration*, the convergence of traditional collaboration, social media, and crowdsourcing, is facilitating brainstorming, enhancing system design, and fostering the launch of new ventures. *Experiential design*, the convergence of system modeling, interactive storytelling, visualization, and augmented reality, is encouraging stakeholders in increasing numbers to participate in evolving system architectures and designs in up-front concept engineering. *Interactive storytelling*, the convergence of engineering and the entertainment/cinematic arts, is transforming engineering design and education. *Interactive novel option generation*, the convergence of psychological principles, systems thinking, and decision science, is enhancing the definition and analysis of trade-off spaces ("trade-spaces"). These examples are covered in depth in the remaining chapters.

3.3 Overcoming System Modeling Challenges

Increase in the number of interconnections and interdependences among system components is a source of ever-increasing system complexity. At the same time, advances in computing and communication are creating a steady shift toward cyber-physical systems. Along with this trend comes the need for precise control of timing to assure safety and correctness [5]. This trend poses a challenge to system modeling, analysis, and complexity management because the intellectual and technological basis of current approaches tend to be rooted in models and tools, including APIs, high-level languages, and virtualization that abstract away time! Therefore, there is a pressing need to develop new, holistic approaches to model, analyze, influence, and shape complex system behavior. As important, there is a pressing need to develop new approaches to manage complex sociotechnical systems and their evolution. A reasonable starting point for understanding complexity is to consider at least five properties that are common to natural, biological, and social systems [4]: nonlinearity, scale dependence, recursive symmetries between scale levels, sensitivity to initial conditions, and emergence resulting from multiple feedback loops and interaction chains. Engineered complex systems exhibit other key properties such as drift into failure and interacting with the protective structure that is provided to prevent them from failing.

3.4 Transdisciplinary Systems Engineering

Applications of disciplinary convergence are diverse and broad. They range from autonomous vehicles and smart manufacturing to integrated healthcare and ubiquitous education. Convergence cuts across cognitive, biological, and social sciences, as well as philosophy (ethics), policy, and culture. The history of science and technology (S&T) shows that a period of divergence typically follows a period of convergence, in which knowledge gained during convergence diffuses to new problem domains. The real value of the convergence-divergence cycle lies in the potential breakthroughs and novel solutions that result from knowledge diffusion.

Convergence is potentially a key enabler of solutions to NAE's Grand Challenge problems. Several major universities such as MIT, Harvard, Stanford, USC, University of Wisconsin-Madison, and the State University of New York are pursuing a variety of initiatives to foster convergence research. Convergence initiatives are also underway in the policy arena in the European Union, Russia, China, Japan, and Korea. In the United States, convergence of biology, engineering, and physical sciences is an active area in the fight against cancer [2]. Disciplinary convergence, the unification of insights and approaches from different fields, will make fundamental contributions in our drive to provide creative solutions to the most vexing problems facing our society. Ultimately, disciplinary convergence is an enabler of *transdisciplinary systems engineering*.

I define transdisciplinary systems engineering as an integrative meta-discipline that reaches beyond systems engineering to make connections with other disciplines with a view to developing solutions to problems that appear intractable when viewed solely through the lens of traditional systems engineering [3]. Examples of other disciplines are social science, cognitive science, biology, economics, philosophy, and culture. In this book, I focus on technology-enabled disciplinary convergence as an enabler of transdisciplinary systems engineering. In other words, transdisciplinary systems engineering is concerned with exploiting the growing convergence between engineering and disciplines from the science and humanities to solve complex scientific and societal problems. Transdisciplinarity is defined as reaching beyond disciplines to find and exploit connections to solve complex problems [3]. Transdisciplinary thinking encourages thinking in terms of new relationships among traditionally distinct disciplines, and focusing on new concepts that might arise from such thinking.

The impetus for transdisciplinary systems engineering stems in part from the growing awareness that traditional systems engineering methods, processes, and tools based on reductive methods are ill-suited for analyzing and designing complex systems. This is because complex systems tend to exhibit unpredictable behavior arising from unknown interactions and dependencies among system elements. This recognition has spurred interest in the development of new paradigms, methods, and tools to: explore complex system behaviors and design trade-offs; detect, diagnose, and visualize component and subsystem interactions; and identify, alert, and circumvent potentially undesirable interactions. In particular, there is a need for a cross-disciplinary framework as the basis for advancing transdisciplinary systems engineering. Such a framework would support modeling, exploration, visualization, analysis, and management of complex systems subject to internal malfunctions and external disruptions.

3.5 Summary

The twenty-first century is being hailed as the era of convergence. We have seen impressive advances in technology convergence with the advent of smartphones and smart homes. We are now entering an era of disciplinary convergence enabled in part by new and emerging technologies. We have already seen advances in Nanotechnology, Biotechnology, Information Technology, and Cognitive Science (NBIC). Now we are beginning to see convergence, at higher levels of integration. As disciplinary convergence continues, it becomes both an enabler of and catalyst for transdisciplinary systems engineering. In this chapter, I have defined transdisciplinary systems engineering as an integrative meta-discipline that reaches beyond systems engineering to seek connections with other disciplines and develop solutions to problems that appear intractable when viewed solely through an engineering lens.

Transforming an institute of higher learning to pursue convergence research and education within a self-sustaining ecosystem is a daunting challenge. It requires overcoming several barriers including the lack of guidance on how to set up an effective program in convergence and the lack of awareness of strategies being employed by other organizations pursuing convergence research. In this regard, the National Academies report summarizes the strategies and practices used by various institutions to facilitate convergence initiatives. The two most significant strategies that this report recommends to facilitate convergence research and education are to organize around a common theme, problem, or scientific challenge and then design education and training programs that foster convergence. The report emphasizes people, organization, infrastructure, and culture as the critical elements in creating a successful and sustainable ecosystem for convergence research and education. Investments in these convergence enablers are critical to making progress on both scientific and societal problems of national and global significance.

References

1. Convergence: Facilitating Transdisciplinary Integration of Life Sciences, Physical Sciences, Engineering, and Beyond. National Academy of Science, Committee on Key Challenges Areas for Convergence and Health; board on Life Sciences; Division on Earth and Life Studies, National Research Council, 2014.
2. Bainbridge W, Roco MC. Handbook of science and technology convergence. New York: Springer; 2016.
3. Madni AM. Transdisciplinarity: reaching beyond disciplines to find connections. J Integr Des Process Sci. 2007;11(1):1–11.
4. Roco MC, Bainbridge WS. Converging technologies for improving human performance: nanotechnology, biotechnology, information technology and cognitive science. New York: Springer; 2003.
5. Alkhatib H, Faraboschi P, Frachtenberg E, Kasahara H, Lange D, Laplante P, et al. IEEE CS 2022 Report. IEEE Computer Society 2014:25–27.

Chapter 4
Disruptive Collaboration

"Many ideas grow better when transplanted into another mind than in the one where they sprung up."
– Oliver Wendell Holmes, Sr., American jurist

4.1 Impact of Innovation on Collaboration

Computer-based collaboration was a natural consequence of networked computers with collaboration software. Collaboration enables brainstorming and idea generation, as well as policy development, product development, and service delivery. The process of collaboration has gone through multiple iterations with each new wave of technology. Social networks, the most recent wave of technology, have disrupted and radically transformed collaboration including the very definition of collaborators. Facebook and Twitter have become an integral part of our daily lives. However, collaboration for the sake of collaboration is not enough. It must be followed through in terms of actions and decisions. In the June 2013 issue of *Inc.* magazine, Eric Paley [1] wrote a provocative essay called "A Great Idea is Never Enough." Paley's refrain that "The system vastly overvalues great ideas and undervalues execution" is on point for the times. Disruptive collaboration enabled largely by social media has transformed how people socialize, work, and learn. It also influences the economics of what societies create and consume. Changes introduced in the way people collaborate tend to be long-term disruptive trends that change the very fabric of societies. Everything from idea generation to product development, manufacturing, and supply chain integration and management is impacted by disruptive collaboration technology. Disruptive collaboration is redefining the germination of ideas and how organizations approach collaboration to fuel innovation. With today's advances in social networks and their use in crowdsourcing, the collaborate-innovate-field cycle can now reach out and touch every individual with access to the Internet.

A.M. Madni, *Transdisciplinary Systems Engineering*,
DOI 10.1007/978-3-319-62184-5_4

4.2 Large-Scale Disruptive Collaboration

Large-scale disruptive collaboration is characterized by a large number of people working together to develop new ideas that change business models, revenue sources, product trajectories, and technology roadmaps [2]. Large-scale disruptive collaboration invariably implies a "paradigm shift." For example, cloud computing is a shift from an in-house IT infrastructure to a "purchase by the yard" service. Similarly, agile development using software as a service (SaaS) has produced a shift from in-house software development to leveraging external SaaS capabilities, i.e., software is bought from a third party and the in-house crew provide the "glue." Clearly, this is the kind of thinking that is needed today to exploit convergence and hyper-connectivity.

Today the rapid proliferation and adoption of social media and crowdsourcing have dramatically changed the very nature of collaboration. In particular, crowdsourcing has become a game changer. Crowdsourcing is a type of outsourcing to contributors outside the traditional network of collaborators to fill a competence gap, or multiply the generation of good ideas and solutions during planning and problem solving. Therefore, crowdsourcing can be viewed as a distributed problem-solving model. Not surprisingly, crowdsourcing is being adopted as a key strategy in software development projects. Specifically, any software development or support activity (e.g., requirements generation, early design, detailed design, implementation, testing and evaluation, documentation) can be crowdsourced to fill a competency gap, accelerate schedule, or reduce costs.

Collaboration today is open and spontaneous and without personal knowledge of collaborators and with a wide variety of social media options (Table 4.1).

Collaboration platforms are a category of business software that adds broad social networking to work processes. The goal of a collaboration software application is to foster innovation by incorporating knowledge management into business processes so employees can efficiently share information and solve business problems more efficiently. Collaboration platforms today offer capabilities for problem exploration, "what-if" experimentation, and solution implementation (Table 4.2).

Traditional Teams Are No Longer the Sole Source of Ideas Today traditional teams are no longer the primary source of idea generation. Crowdsourcing, expert sourcing, and open innovation have become the basis for creating and evolving new ideas, exploring their value proposition, playing "what-if" games with the ideas, and finally implementing them. The concept of open innovation, attributed to Chesbrough of the University of California, Berkeley, has created a true paradigm shift. Ideas and concepts do not come from in-house anymore; rather, they are the collaborative creation with "strangers" for all practical purposes. The same is true of crowdsourcing. Thus, today we buy/license inventions and processes. Also, we expose our own inventions through licensing and spin-offs. For example, InnoCentive is a global web community for open innovation. It helps clients engage a world of creative and diverse on-demand talent to rapidly generate novel ideas and solve

Table 4.1 Collaboration options

• LinkedIn and Facebook-like social media – LinkedIn is a business and professional networking – Facebook is an online social networking service (photos, videos, blogs, apps)
• Twitter – Online social networking and microblogging service that enables users to send and read 140-character text messages, called "tweets" – Microblogging, RSS, updates
• Chatter – Share ideas and documents through enterprise social network – Productive business collaboration
• Skype ScreenShare and GoToMeeting – Skype: Share computer screen as a means to collaborate – GoToMeeting: Collaboration through web conferencing on any platform
• Kickstarter – World's largest funding platform for creative projects in film, music, art, games, comics, design, photography, etc.
• Pinterest – A visual discovery tool to find ideas (e.g., videos) for projects – Create and share visual bookmarks ("pins") to plan, organize, and save
• Flickr – Photo sharing, commenting, photography-related networking, worldwide
• Reddit – An entertainment, social networking service, and news website where registered community members can submit content, such as text posts, or direct links. Only registered users can then vote submissions "up" or "down" to organize the posts and determine their position on the site's pages – Content areas all organized by areas of interests called "subreddits"

important problems. InnoCentive has posted more than 1650 challenge problems to its Global Solver Community which consists of over 300,000 individuals from nearly 200 countries. InnoCentive's customers include commercial (e.g., Procter and Gamble, Eli Lilly), government (e.g., Air Force Research Lab, NASA), and nonprofit organizations (e.g., Lumina Foundation, Rockefeller Foundation).

As important, collaborative teams can be completely virtual (e.g., collaborating over the Internet), face-to-face (e.g., person next door), or hybrid (i.e., a combination of virtual and face-to-face). Figure 4.1 presents a pictorial view of these different types of collaboration.

Today social media has opened up substantially more opportunities for collaboration. Collaboration today is possible among open, distributed, transdisciplinary teams in which collaborators may not necessarily be even aware of each other. In particular, transdisciplinary collaboration is key to opening "mental locks" and breaking down disciplinary barriers during both problem formulation and option generation. Table 4.3 compares and contrasts collaboration within various types of teams.

Table 4.2 Collaboration platform functions

Comparison criteria Functions	Problem exploration	"What-if" experimentation	Solution implementation
Purpose	• Explore problem space • Formulate problem • Link to human solvers	• Generate solution options • Test solution options in relevant contexts • Catalog findings	• Build and test selected solution • Disseminate solution • Solicit feedback
Collaboration strategies	• Assemble and connect stakeholders • Collaboratively surface and resolve issues • Achieve consensus on problem statement • Connect to human/ automated solvers	• Integrate and catalog ideas from stakeholders and human/ automated solvers • Generate test cases to evaluate solutions • Configure environment to conduct testing	• Pursue collaborative solution(s) • Development of solution(s) • Develop solution diffusion strategies • Guide adopters driving implementation • Help manage change cascades
Expected outcome/ desired end state	• Mutually accepted problem definition • Identification of solvers (humans/ automated algorithms) • Candidate solution options	• Evaluation of candidate solutions with different assumptions and alternate futures • Solution selection • Explicit documentation of associated assumptions	• Implemented solution • Implementation standards • Adoption of innovative solution

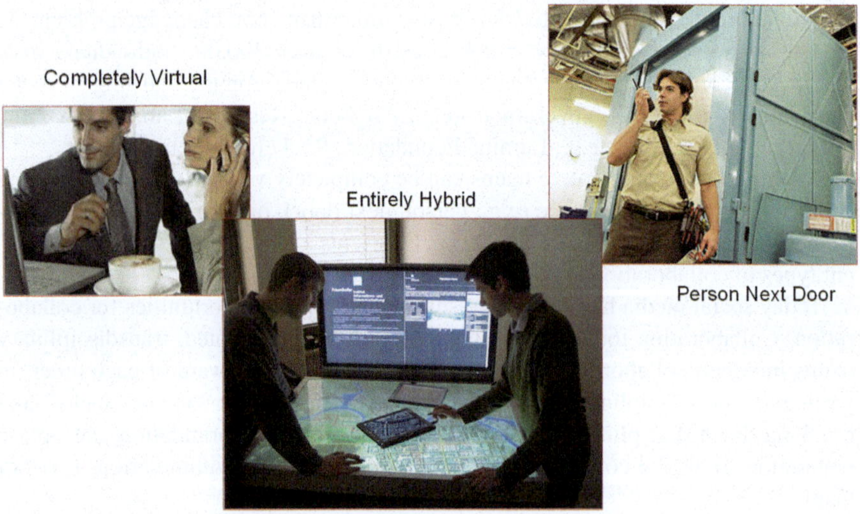

Fig. 4.1 Collaboration options: virtual, face-to-face, hybrid

Table 4.3 Social media provides ample opportunities for collaboration [3]

Research team types / Comparison factors	Intradisciplinary	Multidisciplinary	Interdisciplinary	Transdisciplinary
Collaboration scope	• Among individuals within a discipline	• Among individuals from different disciplines	• Among disciplines through collaborator	• Across and beyond disciplines without regard to disciplinary boundaries
Specific focus	• Deeper understanding within a research field (e.g., quantum physics within physics)	• Achieving compatibility in complex problem solving through collaboration	• Creation of integrative solution potentially resulting in mutual enrichment of disciplines	• Finding hidden connections among knowledge elements from different disciplines
Key characteristics	• Generally, study the same "research objects," (e.g., multiple branches of modern physics) • Tend to have methodologies in common • Tight communications • Mostly speak a common language • Add to the BOK of a branch/ discipline	• Harmonize multiple, occasionally incompatible aspects • Integration limited to linking research results • Susceptible to misunderstanding (specialized languages) • Collaborators occasionally unsure about final resolution	• Development of shared concepts, methods, and epistemologies for explicit information exchange and integration • Can produce an entirely new discipline • Specialization causes knowledge fragmentation, occasionally contradictory knowledge	• Challenge the norm and generate options that appear to violate convention • Look at problems from a discipline-neutral perspective • Employ themes to conduct research and build curricula • Redefine disciplinary boundaries and interfaces

Table 4.4 Unique aspects of transdisciplinary collaboration

• **Goal** – work together to generate and evolve ideas – find creative solutions that transcend disciplinary boundaries
• **Participants** – come together from the very start to exchange ideas – contribute their knowledge and expertise
• **Approaches and solutions** – determined collectively – participants do not develop their own answers to a problem before collaboration begins

In particular, transdisciplinary collaboration can be encouraged in a variety of ways. The most effective way is to ask questions that cut across disciplinary boundaries, because answering such questions encourages participants to reach out to individuals from other disciplines. Transdisciplinary collaboration can be fostered in an environment that encourages "fluidity" and "serendipity." Some of key characteristics of transdisciplinary collaboration are making assumptions explicit to overcome apparent differences, setting constraints aside to foster creative option generation, and actively reaching out to other disciplines to make connections. Some of the key strategies of transdisciplinary collaboration include introducing new metaphors that allow the different stakeholders to get on the same page, changing the level of abstraction to connect with all stakeholders, or sharing a picture or graphic to enhance sensemaking on the part of the collaborators. The collaborative team stays focused on an idea, a problem, or a goal, not on disciplinary expertise. The net result is the dissolving of artificial barriers between disciplines. Finally, large-scale collaboration requires multilayered governance to separate concerns and adhere to scope. The unique features of transdisciplinary collaboration are presented in Table 4.4.

In sum, large-scale disruptive collaboration is ubiquitous and multilayered. It comprises complex combinations of formal and informal networks that range from ad hoc and unbounded to stable and bounded. They exhibit a mix of volunteers and responsible parties. For transdisciplinary collaboration to truly deliver on its promise, there needs to be a blend of creativity and attention to detail during execution and explicit understanding of the key trade-offs.

4.3 Key Trade-Offs

Managing Tensions and Trade-Offs Several trade-offs bear on disruptive collaboration. Some of the more important trade-offs are presented in Table 4.5. Each trade-off is discussed next.

Privacy-Transparency Trade-Off The trade-off between transparency and privacy is quite common. Transparency helps with shared understanding. Privacy protects information that a participant is not willing to share. The key considerations that go into performing this trade-off are trust, need for disclosure, level of disclosure, and

Table 4.5 Typical trade-offs in "disruptive collaboration"

• Trade-off #1: Privacy vs. Transparency
• Trade-off #2: Squandering vs. Withholding Resources
• Trade-off #3: Risk Increase vs. Decrease by Going Virtual
• Trade-off #4: Governance vs. Chaos in Collective Creativity
• Trade-off #5: Stable Leadership vs. Temps in Governance
• Trade-off #6: Platform Design vs. Need for Adaptability

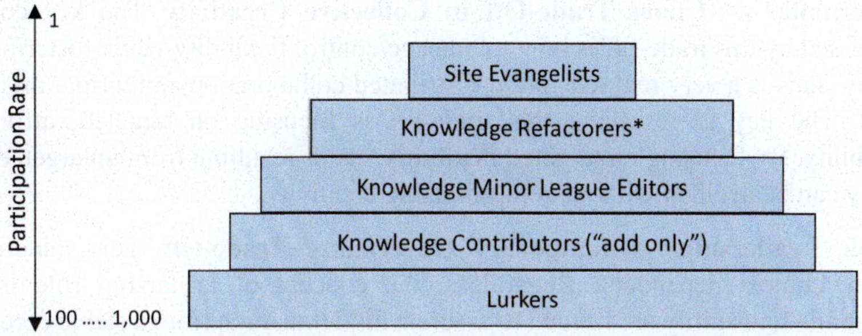

Fig. 4.2 Layers of participation

whether or not an alternate means exists for acquiring the information sought. For example, can one determine the average salary or wage of a group without explicitly having participants provide their salaries or wages. At first blush, it appears that it is not possible to answer this question without asking people's salaries or wages. In reality, it turns out that this question can be answered by using an indirect multiparty computation approach that does not require the participants to divulge their salary or wage information.

Resource Utilization Trade-Off The key here is to avoid resource imbalance. Resources are not just monetary. They can be materiel and social capital also. They also include attention, willingness, and information validation time. Throwing resources at a bad idea or extraneous activity is just as bad as providing inadequate resources for a good idea or needed activity. Thus, the focus needs to be on what resources it takes to evolve a good idea.

Risk Trade-Offs in Virtual Collaboration Virtual collaboration reduces some risks while escalating others. On the one hand, people come together to innovate and collectively lower risk. On the other hand, certain individuals can "shut down" when they have to perform in from of others in the virtual environment. This is because collaboratively innovating is viewed as risky by some individuals. To ameliorate this problem, potential solutions include anonymity in specific contexts and assignment of different roles to collaborators. In light of the foregoing, participants can engage in different levels of participation ranging from being a "lurker" to a "site evangelist." The different layers of participation are shown in Figure 4.2.

The lowest level of participation is that of a "lurker." A lurker deserves the benefit of the collaboration without contributing to the collaboration. The next level of participation is as a “knowledge contributor.” In this capacity, an individual adds to the existing body of knowledge without making any changes to it. The next level of participation is as a “knowledge minor league editor.” At this level of participation, the participant is allowed to and can edit the existing body of knowledge. At the next level are “knowledge refactorers.” Refactoring is the process of rewriting written material to improve its readability or structure, with the explicit purpose of keeping its meaning or behavior. The highest level of participation is the site evangelist who is the most versatile contributor to the collaboration process.

Governance — Chaos Trade-Off in Collective Creativity The key concern expressed by this trade-off is how to manage/control flexibility while fostering creativity. This is a very real tension in distributed collaboration and virtual organizations. The key to resolving this trade-off is focusing on targeted, affordable flexibility. By focusing on targeted flexibility, chaos resulting from untargeted flexibility can be avoided without compromising creativity.

Stable Leadership — Temporary Governance Trade-Off This trade-off is highly context dependent. Regardless, the concept of leadership role is key. Flexibility and agility are needed to assure a fluid transition from a stable core leadership to organic volunteers temporarily performing in governance roles. To accomplish this trade-off effectively, several related strategies can be pursued. These strategies include disaggregating leadership roles, allocating leadership characteristics to those roles, and assigning agents with specific leadership characteristics to these roles, thereby increasing overall flexibility. This flexibility is key to shifting from stable leadership to temporary governance based on context.

Platform Design — Adaptability Trade-Offs This is a key trade-off and one that is becoming increasingly important as the focus shifts to adaptable and resilient systems. On the one hand, platform design standardizes development and reduces development risks. On the other hand, an over-specified platform will tend to suffer from a lack of evolvability and may have to be discarded. Finding the “sweet spot” is a challenge and, in fact, a research problem of great interest. A promising approach to dealing with this trade-off is to selectively incorporate real options in platform design to exploit potential technical breakthroughs without increasing development risks. Figure 4.3 presents a temporal perspective on product platform generation and evolution.

This figure shows the creation and evolution of the product platform. The first box shows the original product platform. The original product platform comprises Generation 1 of the product family along with derivative products 1, 2, and 3. The second box shows an extension to the product platform, which is Generation 2 of the product family. This extension has derivative products 1, 2, and 3. The third box shows the new product platform, which is Generation 3 of the product family. This platform has derivative products 1, 2, and 3.

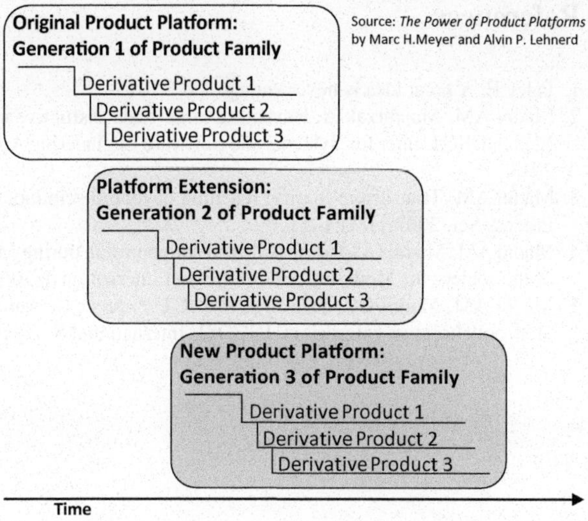

Fig. 4.3 Product platform generation and evolution [2]

4.4 Summary

This chapter has taken a critical look at how convergence between distributed collaboration, social media, and crowdsourcing has led to the emergence of large-scale disruptive collaboration. It has explored the impact of social media and crowdsourcing on distributed collaboration. It has shown that the terms "collaboration" and "collaborators" need to be redefined in light of this convergence. Transdisciplinary collaboration is defined as a means to break down disciplinary barriers and collectively arrive at a solution. Six distinct complexity-driven trade-offs are presented along with effective strategies for performing them. Several provocative conclusions associated with large-scale disruptive collaboration are presented. First, collaborative teams need to focus on idea generation and evolution enabled by technology, people, and organization and not view each factor in isolation. Second, ideas should be adopted as the unit of analysis, with the focus not being exclusively on resolving conflicts among collaborators. Third, context awareness should be exploited to rapidly evolve ideas, without imposing artificial constraints that end up prematurely pruning good ideas and options [4, 5]. Fourth, the focus should be on collaborative behavior, not virtual organization structure. Fifth, organizations should be viewed as organisms with attributes (e.g., people, culture, motivation) that can be exploited, not as constraints to conform to. Finally, collaborative teams should focus on maintaining requisite variety over time, not just pursue satisfaction of "success criteria."

References

1. Paley E. A great idea is never enough INC. 2013.
2. Madni AM, Majchrzak A. Revisiting large-scale disruptive collaboration in the age of Social Media. USC Center for Systems and Software Engineering Annual Research Review; March 14 2013.
3. Madni AM. Transdisciplinarity: reaching beyond disciplines to find connections. J Integr Des Process Sci. 2007;11(1):1–11.
4. Madni AM, Madni CC. Context-driven collaboration during Mobile C2 operations. Proceedings of the society for Modeling and Simulation International. 2004:18–22.
5. Madni AM, Madni CC, Salasin J. ProACT™: process-aware zero latency system for distributed, collaborative enterprises. INCOSE International Symposium. 2002;12(1):783–90.

Chapter 5
From Models to Stories

"Telling purposeful stories is interactive. It's not a monolog. Ultimately, purposeful tellers must surrender control of their stories, creating a gap for the listener(s) to willingly cross in order to take ownership."

– Peter Guber, entrepreneur, educator, author

5.1 Introduction

As systems continue to grow in scope, scale, and complexity, the ability to model, analyze, and design them has become a critical systems engineering challenge. Over the past decade and a half, several modeling languages (e.g., SysML, OPM) have been developed and employed for modeling and analyzing complex systems. These languages require familiarity with engineering notation on the part of stakeholders. Unfamiliar with these modeling notations, the majority of nontechnical stakeholders are unable to contribute to upfront engineering. Today, there is an even bigger challenge given that systems need to adapt to changing operational environments and new regulations, while having the requisite flexibility to seamlessly integrate new, high-payoff technologies.

Current tools are also deficient in addressing the needs of complex system design. These deficiencies tend to dramatically increase engineering change requests (ECRs) and design iterations, inevitably producing schedule delays and cost overruns. In light of the foregoing, new systems engineering tools are needed that address these challenges. Specifically, the new systems engineering tools need to address several key requirements:

- Participation of all stakeholders in upfront engineering and throughout the system life cycle
- Comprehensive definition and cost-effective exploration of solution trade-offs
- Flexible modeling construct that allows system models to reflect changing operational environments and new regulations/standards
- Rapid, cost-effective discovery of latent interactions between the system and the operational environment and assessment of their impact on desired outcomes
- Flexible interfaces to opportunistically and seamlessly integrate promising new technologies

A.M. Madni, *Transdisciplinary Systems Engineering*,
DOI 10.1007/978-3-319-62184-5_5

- Visualization and analysis of how change propagates in a complex system over time and space
- Rapid, cost-effective identification of system "hotspots"

Today, there is a growing recognition that a different approach is needed to communicate evolving system designs and concept of operations (CONOPS) to those outside the engineering discipline. This recognition provides the motivation for exploring a storytelling as a means to engage all stakeholders and facilitate collaborative decision making on their part.

Storytelling has been part of societal and cultural lore throughout history. It bonds the storyteller and the audience in ways not achievable by other means. It fosters understanding and enhances recall by contextualizing key facts and concepts. Storytelling continues to be an effective means for communicating social and cultural norms in virtually every culture and society. Today, it has become a preferred means to engage, inform, educate, entertain, and persuade. From time immemorial, stories have had the same attention-getting effect on young and old from one generation to the next. In general, studies find that when people listen to stories, they experience a feeling of being "transported" to an alternate world. Interestingly, when the story concludes and the audience "returns" to the real world, they often do so with altered beliefs. In fact, there is evidence from psychological studies that stories can leave lasting impressions and produce durative belief change [1]. Not surprisingly, people view stories in the same way they view logic, when it comes to serving as evidence for beliefs. Visual and multimedia storytelling in virtual worlds can create an even greater impact.

Storytelling has made dramatic strides over the last decade. Diverse domains such as sales and marketing, patient-clinician engagement, disaster management, and education and training are harnessing the persuasive powers of effective storytelling. More recently, systems engineering has begun to exploit the immersive powers of storytelling to convey an evolving system design and concept of operations (CONOPS) to all stakeholders, many without the requisite technical background to use system modeling languages such as SysML [6]. Stories, represented using formal languages and enacted within virtual worlds, can dramatically enhance collaboration among all stakeholders, while also circumventing potential confusion that arises from the use of free-form, unconstrained text to describe operational scenarios and system goals.

Today, storytelling in virtual worlds is being used for system design and CONOPS development (Table 5.1).

A virtual world looks and feels like the real world but without the physical constraints of the real world. In virtual worlds, avatars represent people and can perform real-world activities. Virtual worlds blur the boundaries between art and science, technology and cinema, and biology and technology [2–4]. This blurring of boundaries fosters imaginative thinking. The virtual world can serve as a platform and environment for systems engineering in which system stories can be explored with different assumptions over different time horizons. This recognition provided the impetus for mapping engineering models to system stories for execution in virtual worlds. This capability allows stakeholders to interact with the system (model) and

Table 5.1 Storytelling in virtual worlds

• Creation of a social experience in which culture and language harmonize
• Characters and viewpoints are harmonized within the story environment
• Principles from cognitive and social sciences are used to inform behaviors of characters/teams
• Extensibility/evolvability and scalability are key drivers
• Technology and the creative process are harmonized
• Story ontology enables mapping and contextualizing of system design and CONOPS in the virtual world
• Story developers engage in open, virtual collaboration in the real world to specify virtual world entities (i.e., design, environment, characters, and viewpoints)

explore system behavior within the context of system stories that are interactively executed in virtual worlds. This is interactive storytelling.

Interactive storytelling is a form of digital entertainment in which the storyline is not predetermined. The story author sets the context, the characters, and their roles that the narrative addresses. Users can experience unique stories based on their interactions with the story in virtual worlds. Interactive storytelling cuts across several disciplines including psychology, sociology, cognitive science, linguistics, natural language processing, user interface design, computer science, and computations intelligence. Interactive storytelling allows users to explore alternate story lines in virtual worlds. By changing assumptions or inserting "injects" into the story, stakeholders are able to systematically explore system behaviors and outcomes in a variety of alternate futures.

5.2 Model-Based Interactive Storytelling (MBIS)

Over the years I have come to believe that engineering models need to be "wrapped" and packaged as stories to connect with all stakeholders (i.e., not just technical stakeholders). Thus, over the past few years, I have developed a model-based interactive storytelling (MBIS) approach that uses system models produced by model-based systems engineering (MBSE) in conjunction with operational sequence diagrams and use case scenarios to construct system stories. In essence, MBIS exploits the growing convergence between engineering and entertainment arts. Specifically, MBIS combines MBSE and interactive storytelling in virtual worlds. With this approach, stakeholders are able to interact with partially authored system stories in virtual worlds to explore system behaviors with different technologies, assumptions, and initial conditions. This process enables discovery of previously unknown interactions and dependencies among system elements and between the system and the environment. The ability to interact with stories in virtual worlds significantly enhances requirements definition and trade-off analysis performed in upfront engineering. By involving all stakeholders through system stories, this approach potentially avoids extraneous design iterations and rework, which have historically led to program cost overruns and schedule delays.

Surprisingly, much of what is known about the cognitive effects of storytelling has gone untapped thus far [5]. Thus, a productive line of inquiry consists of:

(a) Assessing the degree to which MBIS can, in fact, achieve increased understanding of system-related issues and lead to stakeholders' buy-in.
(b) Creating and exploiting an MBIS framework that promotes increased understanding and acceptance of system design and change proposals.

MBIS harmonizes the rigor of models with the innate appeal of stories [6]. A system story, constructed from a combination of system models, use cases, operational scenarios, and the knowledge acquired from subject matter experts, is the primary vehicle for analyzing system behavior. A story consists of a combination of multiple vignettes. In MBIS, MBSE-generated models are mapped to a series of story vignettes (i.e., narrative use cases), which bring out the salient aspects of system behavior during interactive story execution. Vignettes consist of nominal vignettes and contingency vignettes. Users can explore alternate system behaviors by substituting contingency vignettes for nominal vignettes or injecting disruptive vignettes at appropriate points in the story. In some sense, MBIS is analogous to model-based testing. Instead of writing tests to exercise code, we write stories to exercise system designs. An apt analogy is comparing a model-driven story to a melodious song. Stories are to models, what melody is to lyrics. We pick up the melody long before we pick up the lyrics. Similarly, we relate to stories long before we understand models.

As noted earlier, MBIS maps system models to system stories. The combination of systems modeling and storytelling within virtual worlds has the potential of transforming how system designs and CONOPS are envisioned and communicated to stakeholders [6]. It is possible to communicate systems engineering artifacts created during system modeling in nontechnical terms in the virtual world. Virtual worlds offer cost-effective and convenient storytelling platforms for assessing system designs and CONOPS development. Today, it is becoming increasingly possible for collaborating stakeholders to explore the structure and behavior of an evolving system designs and associated CONOPS within stories that can be interactiely executed in virtual worlds. Specifically, system models, which are transformed into system stories, allow all stakeholders to explore system behavior from their respective perspectives. The pillars of interactive storytelling are:

- *Systems engineers*, in the role of authors, partially script stories around the system model and CONOPS; the rest of the story results from interactive story execution in the virtual world.
- *Stakeholders*, including customers and users, interact with the system (model) to explore its behavior within the story context using a variety of "what-if" changes to assumptions, operational conditions, and system properties.
- *System models*, from MBSE, are mapped to system stories with model and story entities being represented in virtual worlds.

- *System stories*, constructed from system models represented in SysML or some other system modeling notation.
- *Virtual worlds*, the multimedia computational environment within which stories are interactively executed by stakeholders.

MBIS has the added benefit of allowing stakeholders to interact with the system model and in so doing steer the story in desired directions of interest. In other words, stakeholders can influence the evolution of the story as they seek answers to specific questions. Consequently, stakeholder interactions with model-based stories can be both engaging and illuminating at the same time [7, 8]. For MBIS to fully deliver on its promise as a preferred method for complex systems engineering, two key advances need to occur. First, systems engineers (typically nonprogrammers) need to be able to author system-related stories from the perspective of different stakeholders. Second, systems engineers need to be able to employ appropriate metaphors and methods to convey technical aspects of a complex system in nontechnical terms using multimedia technology. The latter is essential for developing a shared understanding of system behaviors among stakeholders. In turn, shared understanding is essential for reaching consensus on desired system functionality, features, and behaviors.

5.3 MBIS Enables Transdisciplinary Systems Engineering

MBIS, an example of disciplinary convergence, is an enabler of transdisciplinary systems engineering [9]. Its building blocks are *model-based systems engineering*, *experiential interfaces*, and *interactive storytelling in virtual worlds*. MBIS exploits MBSE and interactive storytelling to construct system stories [6, 7]. System stories bring out the key aspects of system behavior. In MBIS, the stories are about the system and its behavior with different assumptions and in a wide variety of circumstances. Each building block is discussed next.

Model-based systems engineering (MBSE) calls for a shift from document-centric to model-centric engineering and the development of a formal system modeling language such as SysML [6] and, more recently, OPM. MBSE proponents stress the importance of modeling to ensure that all key requirements are addressed systematically and rigorously. SysML supports both object and process modeling, with several different types of diagrams (Table 5.2).

Today, SysML is the preferred language for MBSE. It can be specialized for different system domains, such as smart manufacturing and healthcare services. Systems engineers have used SysML to successfully model organizational and enterprise structures and behaviors. However, when it comes to modeling sociotechnical systems, SysML has certain known limitations. It offers the use case diagram to represent system usage by different stakeholders. However, to represent behavioral and social phenomena, it employs state transition diagrams, which are quite limited.

Table 5.2 System diagrams

• Requirements diagrams	
	– to model requirements, their interrelationships, and relationships with other system elements
• Structure diagrams	
	– block definition diagrams to model system structure
	– internal block diagram to represent interfaces and interconnections within a block
• Behavior diagrams	
	– activity diagrams to model state-based behavior (perspective of inputs, outputs, controls)
	– sequence diagrams to model sequences of events/messages involving different system elements
	– state machine diagrams to model system behavior from perspective of state changes caused by events
	– use case diagrams to represent users interacting with a system and desired outcomes
• Parametric diagrams	
	– to represent constraints on system parameter values
• Package diagrams	
	– to organize various model elements (similar to UML)

To overcome this limitation, several researchers are working on incorporating more sophisticated human models within the MBSE rubric [10].

Experiential design interfaces or experiential interfaces (for short) exploit intuitively appealing abstractions, metaphors, and visualization schemes in communicating with users. With regard to MBSE, an experiential perspective augments standard system model representations in SysML (or some other system modeling notation) with views that enable exploration of system behaviors within stories that are interactively executed in virtual worlds [6]. Experiential perspectives place the design within its operational context. The operational context is defined by partially scripted stories that stakeholders interact with to explore system behavior in virtual worlds. Stories comprise events, timelines, system models, geo-located objects, environmental entities, and their behaviors (if applicable). To create experiential perspectives, the role of the system designer/engineer is expanded to include that of a creator of experiences that specifically address stakeholder concerns [6].

When employing MBIS, systems engineers "wrap" partially scripted stories around the system to contextualize the system in its operational environment (i.e., virtual world). The story comprises partial scripting, key events, and "way points" to maintain coherence in the story as stakeholders (users) interact with the system within the story to explore system behavior from their respective perspectives. These interactions result in alternate futures that collectively illuminate deficiencies in system design and CONOPS. Thus, story control is shared between story authors (scripted part of the story) and stakeholder users (interactive part of the story). The latter can become actors in the story when they interact with the system (model) in virtual worlds.

Table 5.3 Mapping between design space and storytelling space

Model	Virtual world – dynamic representation within which story unfolds, system behavior is illuminated as subsystems interact and system interacts with environment, characters interact, and their collective interactions alter the story trajectory and mission outcomes
Use Case	Story – illustrates how system is used, by whom, and under what conditions; story contextualizes use case(s) allowing stakeholders to critique system behaviors and collaborate on needed system changes
Actor	Character – participant within story or narrator of story; have personalities and preferences that influence how they interact with the system and other story characters; may be a SME interacting with the virtual world or be a part of the virtual world
Viewpoint	Author-defined viewpoint – story enacted from author-defined point of view
Basic Flow of Events	Main storyline – plot defines events, characters, and interactions between among them
Alternate Flow of Events	Vignettes – multiple clusters of interactions among specific characters who interact with the system as story unfolds
Exception Flow of Events	Surprise vignettes – represent twists in storyline and/or unexpected or surprising ending resulting from what-if changes to system properties and resulting system behavior; help illuminate system resilience to unanticipated and emergent scenario conditions
Association	Character interaction – defines which characters interact with system under what contexts
Dependencies	Character behavior information dependencies – define information that the system needs to provide to characters to perform their respective tasks
Composition	Family tree – defines how subsystems relate to each other and to the characters in the story

Stakeholder experiences can be shaped by the multisensory environment and the use of the right metaphors (e.g., structural, orientational, ontological). The *structural metaphor* presents one concept in terms of another (e.g., frame an argument as war). The *orientational metaphor* pertains to structuring experiences in terms of spatial directions (e.g., down direction is negative, and up direction is positive). The *ontological metaphor* helps structure experiences of abstract phenomena in terms of concrete objects and forces (e.g., risk level in terms of colors, proximity in terms of symbol size, interaction strength in terms of color and link thickness). From a systems modeling and engineering perspective, this means that system designers introduce conditions along with the system model that contribute to stakeholders' experiences when they interact with the system model within the stories in virtual worlds.

The role of the experiential perspective goes beyond understanding and communication to being a valuable source of experience and insights during collaboration. The experiential perspective can also capture how stakeholders describe their experiences with the system when stories are interactively executed in virtual worlds.

Stories provide a convenient basis for contextualizing systems engineering artifacts associated with system design and CONOPS [11]. Table 5.3 presents the concept of mapping systems engineering artifacts from the design space to technical story artifacts in the virtual world. The systems engineering artifacts include key entities associated with system CONOPS and design, the relationships among these entities, their structure and behavior, human roles in the system, environmental factors, and viewpoints.

MBIS employs role-specific and function-specific "lenses" to contextualize information for the different stakeholders. The more contextualized knowledge is for the different stakeholders, the greater the impact that knowledge will have in terms of shaping their understanding and influencing their beliefs [11–13]. Contextualization can be challenging because it is not practical to have human authors manually create customized story content for every stakeholder user. Therefore, to avoid the manual process, artificial intelligence techniques can be employed that allow the software to assume partial responsibility for content creation and contextualization [14]. An intelligent agent reasons about the most effective way to change the storyline/experience to convey specific system insights to stakeholder users, to enable them to draw well-founded conclusions [7, 11]. Contextualization, achieved through relatively minor adjustments to the model-based story content, is realizable with current technologies such as customizable templates. The underlying rationale is that, unlike dramatic stories, model-based stories that illuminate specific aspects of system behavior are more purpose-driven and system behavior-oriented. As such, many of the computational models of cognition and goal-driven behavior apply to stories [15]. Approaches to contextualization range from parameter adjustments and customizable templates to narrative generation [1]. Allowing stakeholders to choose "lenses" and "filters" when viewing and interacting with system stories increases system contextualization and stakeholder understanding.

5.4 Computational Representation of a Story

For a story to be executed and adapted on-the-fly, it must be represented in a form that lends itself to manipulation by reasoning algorithms [15]. An example of a suitable representation for narratives is *partially ordered plans*. A plan contains steps (i.e., events that change the state of the world) and annotations (i.e., explicit marking of the temporal and causal relationships between the steps in a plan) that define a partial order, which informs the execution sequence of steps. Plan-based representations have the advantage of being able to treat a set of required features for a new narrative as goals and being able to relate all generated narrative elements back to those goals. That is, there is a clear relationship between what happens in the narrative and user requirements. An additional benefit to employing the planning construct for narrative representation is the ability to reason about causality. Work in narrative understanding shows that causally related events form a more cohesive narrative and are thus easier to understand and reason about [15, 16].

The representation of a story in computational form is needed to manipulate and customize stories [7, 11, 15]. Structural narratology, i.e., the study of narrative from a structural perspective, provides some insights into how stories can be represented

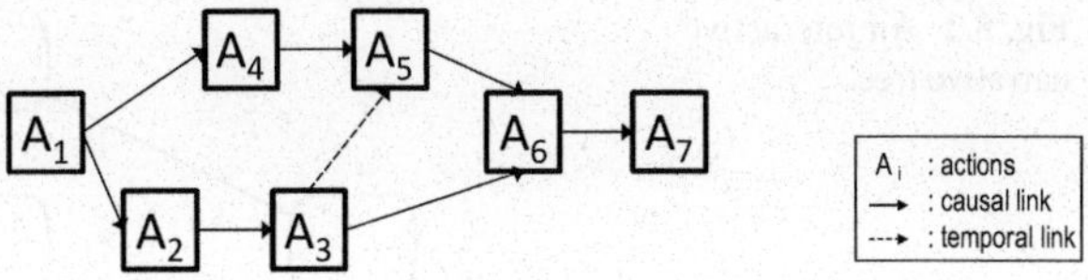

Fig. 5.1 A narrative plan graph

computationally [1, 16]. Structural narratology suggests that it is possible to analyze a narrative at different levels: *fabula* and *sjuzet*. The fabula is an enumeration of events that occur in the story world from the time that the story begins to the time that it ends. Events in the fabula follow the order of their occurrence, not their specification order. The sjuzet of a narrative is a subset of the fabula. The sjuzet is the narration that the audience hears or sees. If the narrative is written or spoken word, the narration is in natural language. If the narrative is a cinematic presentation, the narration is through the actions of actors and the camera shots that capture that action. While it is the narrated sjuzet that the audience experiences, it is the fabula of a narrative that is the content of the narrative and what the narrative is about. At the level of fabula, it is possible to represent a narrative or story as partially ordered sequence of events, where an event can be an intentional act performed by a story world character, or an unintentional action. Computationally, a graphical representation of narrative resembles a partially ordered plan. Figure 5.1 shows an example of a narrative, represented as an acyclic graph.

In this figure, a directed graph represents a narrative. The nodes in the graph represent actions or events that occur in the world. Actions map one set of world states to another set of world states – actual world state does not have to be explicit, allowing for greater flexibility in planning and execution. Dashed links represent temporal ordering constraints independent of any causal relationship. For example, A3 must temporally precede A5. Solid links represent causal relationships between actions and events – an action establishes a propositional truth in the world state that is necessary (but not always sufficient) for the successive action. The execution order allows for parallel actions or, in the case of discourse, several possible total orderings. The partial order planning formalisms provide the basis for several metrics:

- *Plan correctness*: A sound plan, in the absence of uncertainty, executes correctly because it is possible to guarantee, based on temporally preceding actions (or initial world conditions) and the causal conditions required for correct execution of each action.
- *Story coherence*: A coherent narrative is one in which actions are all part of causal chains that lead to one of the outcomes.
- *System complexity*: The number of unique dependency chains or the average number of causal in-links and out-links for each behavioral action can be used to measure system complexity.
- *Explanatory power*: The basic representation (Fig. 5.1) is extensible with nodes representing character intentions to determine the explanatory power of a story – the degree to which preceding events establish character actions and intentions.

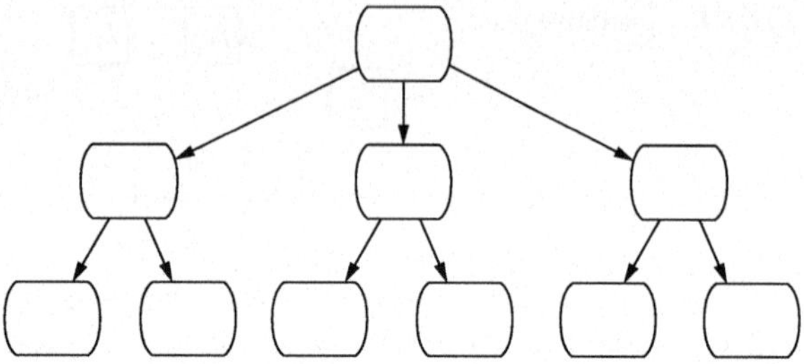

Fig. 5.2 An interactive narrative tree

Plan representation in Fig. 5.1 does not address interactivity, which manifests itself as *branching* of narrative structure. Figure 5.2 presents an interactive narrative tree in which each node represents a fragment of a narrative plan graph.

A suitable representation for an interactive narrative is a tree structure in which each node encapsulates a complete narrative. The root node is the exemplar narrative (i.e., the ideal narrative progression). Links represent conditions under which a narrative is no longer executable – the user performs an action that causes p to be negated during interval i where p must hold during interval i for narrative progression. The remaining nodes are contingency narrative plans that continue the narrative should the associated condition occur (Fig. 5.2). Any path through the tree, involving execution of a prefix of the exemplar narrative plan, a branching condition, the execution of a prefix of a contingency narrative plan, etc., is a complete, coherent, and sound narrative and is provably equivalent to the type of branching structures found in *Choose Your Own Adventure* novels. The advantage of this representation is that it allows for continuous real-time execution such that the user need not be aware of the branching structure.

5.5 Computational Framework for MBIS

A computational framework supports system modeling, story authoring, and interactive story execution in virtual worlds. Figure 5.3 presents the system concept of the model-based interactive storytelling framework. This framework serves as a bridge between system models and story execution in virtual worlds.

As shown in Fig. 5.3, stakeholders interact with partially scripted system stories to explore system behaviors under different conditions from their respective perspectives. Specifically, stakeholders interactively explore potential design alternatives and their implications on their respective "win conditions." For example, the story could be about an individual (or team) working with a complex system. Real-world stakeholders can interact with the system in the story and steer the story in desired directions by making "suggestions" and then observe how the system behaves in response to hypothesized events or conditions. The story presents the system in a fashion that dispels erroneous beliefs and assumptions about system design on the part of the stakeholders while simultaneously reinforcing proper

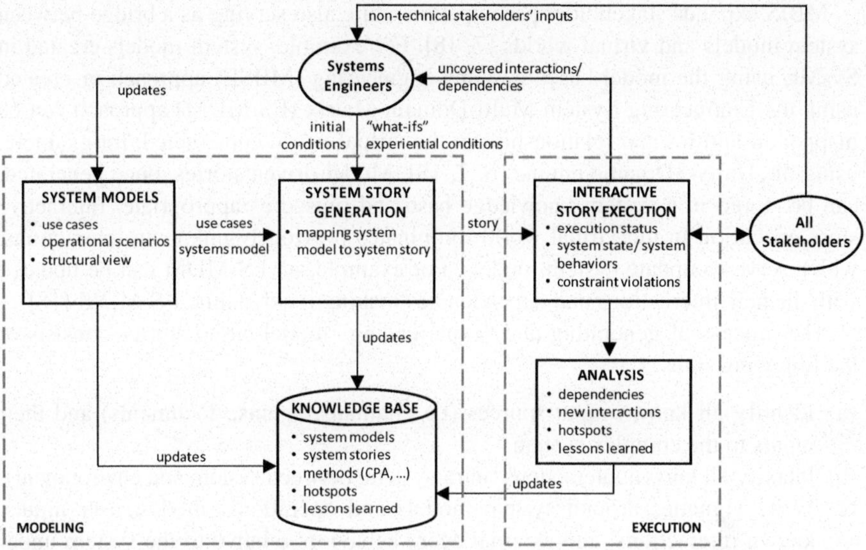

Fig. 5.3 MBIS in virtual worlds

understanding of system behavior. To effectively convey information about system behavior, the storytelling system can employ appropriate metaphors such as structural, orientational, and ontological [17], along with geospatial visualizations to communicate key system aspects such as nonlinear behavior, system sensitivities to specific parameter changes, change propagation paths and change cascades, and hotspots.

The *structural* metaphor expresses one concept in terms of another (e.g., security system structured along the lines of a human immune system, argument structured as war). Similarly, the *orientational* metaphor comes into play when structuring experiences in terms of spatial directions (e.g., down direction is negative or decreasing; up direction is positive or increasing; and sideways direction indicates no change). The *ontological* metaphor helps structure experiences of abstract phenomena in terms of concrete objects (e.g., risk or threat level in terms of colors, proximity in terms of symbol size, size in terms of relative dimensions with respect to a known object, interaction strength in terms of link thickness or color intensity). From a system modeling perspective, this means that systems engineers need to design conditions and introduce cues that enrich stakeholders' experiences and increase their understanding.

MBIS also assists systems engineers in developing system-related stories that can be adapted to other systems and domains with minimal effort. For example, a story created for aircraft health management should be readily adaptable to crew system design by changing the storytelling knowledge base and the multimedia art assets. The approach and the framework essentially remain unchanged.

MBIS expands stakeholder participation while also serving as a bridge between system models and virtual worlds [7, 18]. For example, system models created in SysML using the model-based systems engineering (MBSE) approach or created using the Engineering System Multi-Domain Matrix (ES-MDM) approach can be mapped to entities and relationships in the virtual world constructed, for example, using the Unity 3D game engine [6, 7, 18]. Model-driven stories, once generated, can be stored in a system knowledge base and reused as appropriate. Interactive story execution in the virtual world illuminates previously unknown interactions, which serve to update system models. For example, an ES-MDM can be updated with the new interactions and dependencies making it a dynamic ES-MDM [18].

The process of generating and "experiencing" model-based stories consists of the following steps:

(a) Identify all known input sources (i.e., humans, agents, documents) and their inputs to the complex system.
(b) Identify all known interactions (intra-system, between system and environment).
(c) Write vignettes around system models and operational modes, user inputs, known interactions, and external agent behaviors relative to the system under development.
(d) Embed multimedia cues and conditions into vignettes to draw attention of various stakeholders (i.e., users) to various aspects of the system ("experiential design").
(e) Convey interactions and flows from the perspectives of the different stakeholders (i.e., stakeholder lenses) while allowing them to introduce "what-if" changes into the vignettes.
(f) Allow users to critique system behaviors/responses to various inputs and conditions; the critique can be with respect to various variables (e.g., time delays, response magnitude, undesirable interactions, cost penalties, unclear/cryptic interactions with the users).
(g) Allow users to understand ripple effects and change cascades (intended, unintended) using animation and visualization techniques in response to user-supplied "what-if" changes.

5.6 MBIS Approach

The MBIS approach encompasses six functions – mapping system models to system story elements, development of role-specific and function-specific lenses, user-system interaction prototyping, story authoring and vignette definition, interactive story execution, and execution data collection. Each function is described next.

Mapping System Models to System Story Elements A common practice in defining system goals and stakeholder needs is to create representative operational scenarios (use cases) that systems engineers can use to collaborate with subject matter experts

(SMEs) and nontechnical stakeholders. It is quite common to find operational scenarios documented primarily in prose form in operational scenario documents. These documents tend to have inconsistencies, ambiguities, and gaps. Not surprisingly, systems engineers prefer formal, structured diagrams such as use cases and activity diagrams. However, most SMEs and nontechnical stakeholders tend to be unfamiliar with systems engineering methods and formal notations. Thus, effective collaboration between systems engineers, SMEs, and other nontechnical stakeholders continues to be a major challenge. This recognition provides the impetus for mapping system models to system stories that stakeholders can individually and collaboratively explore. Specifically, system models, use cases, operational sequence diagrams, and SME inputs are used to create system stories. Table 5.3 (presented earlier) provides the mapping from the system design space to the system story space (virtual world). The virtual world is a domain-limited yet verifiable representation of the system of interest. Stakeholders can collaboratively perform model verification.

It is important to note that while several system model elements will map to entities in the virtual world, they will only partially populate the virtual world. System stories, which are informed by use case scenarios, system models, and expert knowledge, are used to supply the remaining entities that populate the virtual world. Story vignettes also provide the context that facilitates the mapping of system model artifacts to virtual world entities. Partially automating the mapping of system elements, events, actors, parameters, and relationships to engaging narratives is, in fact, possible. The storytelling environment allows all stakeholders to explore the what, how, and why of system behavior from their respective perspectives for a variety of conditions and with different assumptions.

Development of Role-Specific and Function-Specific Lenses Role-specific and function-specific lenses are a means to reduce apparent system complexity by making a complex system comprehensible to all stakeholders from their respective perspectives. This feature allows stakeholders to provide meaningful inputs about system behavior from their respective perspectives. Role-specific lenses allow users to view the interactions within and outside the system from the perspective of a particular role. For example, in the aircraft divert scenario, the relevant roles are pilot, flight crew, maintenance controller, ground crew/line maintenance, and airline operations center. Examples of functional lenses are fault history, communications, fault status, prioritized faults, and forwarded faults.

Story Authoring Story authoring requires the systems engineer to also take on the role of a story author. Using a combination of operational scenario use cases, system models, and expert elicitation, stories and associated vignettes are developed. Story vignettes have embedded conditions and cues in the operational scenarios that help illuminate system behaviors and create insightful experiences in virtual worlds. The storyline and each vignette have specific openings (i.e., what is true when the story/vignette begins), triggers for each vignette, and conclusions (i.e., the terminating conditions for each vignette/story). It is important to note that the storyline is not linear and depends on the interactions between the system and agents in the simulated operational environment.

User-System Interaction Prototyping The purpose of user-system interaction prototyping is to: a) create story-driven stakeholder experiences with the system model in virtual worlds; b) provide feedback in the form of findings to system modelers to update/revise the system model; c) partially populate the virtual world with entities and relationships addressed by the limited (i.e., horizontal prototype). The story provides contextual information such as location, environmental entities, terrain, weather, and visibility. Collectively, these information items help create an experiential environment for stakeholders to interact with and develop an understanding of the system. The output of user-system interaction prototyping is a horizontal prototype [19].

Story Execution in Virtual Worlds Story execution in the virtual world is key to engaging all stakeholders in a manner that allows them to understand, interact with, and explore system behavior with different assumptions. Specifically, stakeholders can increase their understanding of system behaviors, explore system behaviors under a variety of "what-if" circumstances, and suggest changes to system design from their own perspective (i.e., lens). The lenses can reflect the perspective of a particular role/function or specific impact factors that the stakeholder is interested in. An example of a role-/function-related factor is "communication." Examples of impact factors for this illustrative example are cost, schedule, risk, airline delay, maintenance turnaround time, airplane diverts, canceled flights, and customer satisfaction. Stakeholder feedback to system modelers is in the form of recommended changes to system behaviors and other key properties (e.g., appearance, interactions with other entities). Stakeholders can also offer opinions on the usability and overall utility of the system.

Interactive Execution of Story Vignette Execution of story vignettes is initially an "interactive step through" the story, in which time-stamped scenes are presented to stakeholders in a particular sequence to create semi-immersive experiences. The behavior of the system becomes apparent to stakeholders as they follow the story sequence. Eventually, automated sequencing of scenes associated with the story can replace human step through. The Unity 3D game engine is an example of the story engine that can execute stories interactively in virtual worlds defined by the Unity 3D environment.

Story Execution Data Collection As stakeholders interactively execute stories in the virtual world, tacit knowledge in the form of hidden interactions becomes explicit. The system model, especially use cases and activity sequences, is updated with the discovery of hidden interactions. The system concept of operations (CONOPS), user-system interactions, and user interface are refined using findings from user-system interaction.

5.7 Overcoming Implementation Challenges

Despite the obvious appeal of stories, storytelling is difficult. There are even fewer qualified "storytellers" (i.e., story content creators capable of effectively delivering a message) when it comes to telling stories to large audiences. While technologies exist that can generate stories from a relatively small set of input knowledge, these technologies are still relatively nascent despite nearly 30 years of research in story generation systems. This is where the opportunity exists. Making stories interactive increases their relevance for individuals because they are able to inject their own inputs about system structure and behavior and receive appropriately changed story content that reflects the impact of their decisions. Recent research in artificial intelligence and *computational creativity* suggests that for reasonably constrained domains, content generation is potentially possible. This is especially true of complex systems engineering in which the goals are relatively well defined. In fact, much of what we know about storytelling from cognitive science and discourse theory applies to complex systems engineering.

A branching storyline geometrically increases content authoring time based on the number of branching points and the number of possible branches per point [11, 16]. That is, for every point at which the user can make a decision, the amount of content produced is a multiple of the decisions made (assuming no loops). Consequently, interactive stories with branching require an exponentially greater amount of content to be produced to cover every possible contingency that can result from user actions and explorations [16]. These facts present an interesting set of challenges. First, systems engineers have to become storytellers. Second, there needs to be an adequate number of systems engineers trained as storytellers to create the requisite content to cover system design-related issues of stakeholders' interest. In addition, there are two key complications with regard to authoring content:

- Graphics and animation – one cannot assume that systems engineers have the requisite artistic abilities for content authoring.
- Authoring branches – one cannot assume that systems engineers without adequate training can set up sufficiently complicated scenarios with branching structures required to step through the necessary permutations of the different situations.

In light of the foregoing, there is a pressing need for intelligent authoring tools [14] with the requisite knowledge of the system domain and storytelling to provide scaffolding assistance to systems engineers during content authoring. In this regard, recent work [15, 16] on AI for story generation and interactive storytelling has made strides in developing computational systems that can represent and reason about story structures. While the ability to perform computations automatically based on computer-encoded knowledge is essential for helping systems engineers author branching stories, there is also the need to understand the typical mistakes that they are likely to make. In systems architecting and design, the ability to reactively and preemptively provide feedback regarding errors is called computer-aided critiquing.

In storytelling, there needs to be positive feedback and reinforcement provided to human authors responsible for building an audience model (i.e., a model of a hypothetical target user or user group). This response to authored story content should be similar to what humans would provide.

Finally, we need to integrate the interactive storytelling tool into the systems engineering social context in which users employ the tool. Tool integration within a social context requires a careful match of the tool to the user's needs, abilities, and social support infrastructure [20]. In addition, there has been extensive research such as channel expansion theory [20] and technology coping theory [21] to identify a host of additional factors that can potentially influence the extent to which a new tool can be effectively used by users. Therefore, we need additional research to "pilot test" and refine ways to optimize this match for the systems engineering domain. More specifically, for an interactive technical storytelling tool to secure stakeholder buy-in, we need to match the following elements:

- The technical storyline, characters, setting, and action with the users' perspectives.
- Behavioral choices offered with the choices that users see as feasible and likely given their respective perspectives.
- The format and content with the skill and educational level of users.
- The message of the story needs to be adapted to each stakeholder's concerns and interests. For example, a stakeholder interested in a highly resilient system will expect a different set of system behaviors from the storyline than one who is interested in just predictable responses.
- The message of the technical story needs to be adapted to each user's perceived control over system behavior and expected outcomes.

5.8 Illustrative Example: Campus Security System (CSS)

This subsection presents MBIS concepts within the context of a campus security system (CSS). The campus security example is used to illustrate the modeling perspectives, the mapping from system models to system stories, and the interactive story perspectives in virtual worlds.

The illustrative example revolves around a CSS's surveillance coverage. The overall purpose of the CSS is to ensure the safety of the campus using its limited resources. The CSS is an automated campus monitoring, surveillance, and response system. It consists of computer-controlled surveillance cameras and sensors that are strategically located in various structures across the campus. The CSS interacts with campus security personnel and communicates with police, fire department, and paramedics (Fig. 5.4).

As shown in the figure, the CSS encompasses surveillance cameras, security monitors, campus police, communication devices such as cell phones, geospatial maps, security personnel, and communication networks. The basic concept of operation

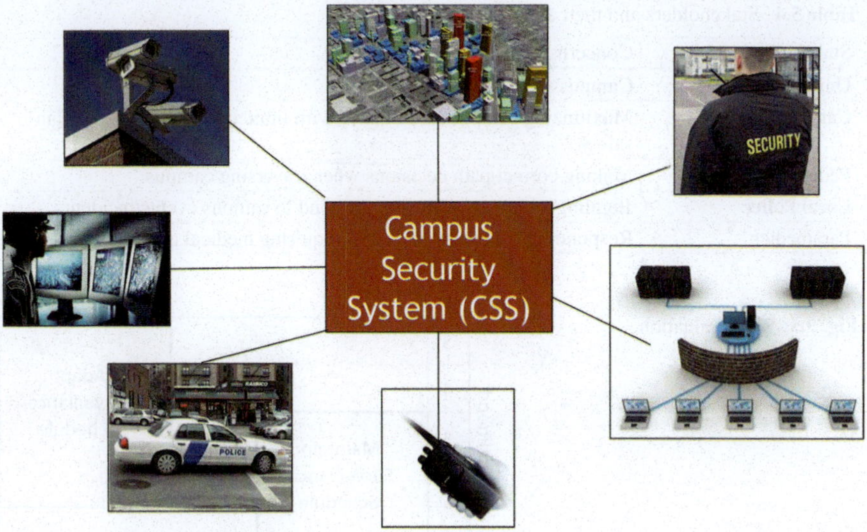

Fig. 5.4 Campus security system

of the CSS is rather straightforward. It receives inputs from sensors, cameras, students, pedestrians, police, and other services. The surveillance coverage is based on known pedestrian traffic patterns on campus as a function of time of day. However, known traffic patterns are disrupted when a special event (e.g., ball games, political fundraisers) is held on the campus or in the immediate vicinity. In these circumstances, the otherwise adequate surveillance schedule becomes inadequate. The question that needs to be answered is whether surveillance coverage should be adapted without adding resources when traffic patterns change in response to special events occur.

We begin by identifying high-level requirements. The high-level requirements for the campus security systems are to ensure campus security and pedestrian safety, respond to campus incidents by dispatching security personnel and alerting police, position cameras and sensors to maximize coverage with limited resources, adapt surveillance schedule in response to special events occurring on campus or in adjacent areas, and alert security personnel in case of malfunction.

The next key step is to identify stakeholders and their areas of concerns. The main stakeholders and their concerns are presented in Table 5.4.

As shown in Table 5.4, the main stakeholders in our example are the university president, campus security, pedestrians, local police, and paramedics.

The problem that the campus faces is that the full value of the CSS is not always realized because the fixed surveillance schedule does not work as well (i.e., there is an increase in campus incidents) when traffic patterns change as a result of a campus event or an event in adjacent areas. The approach to solving this problem is to evaluate the risks associated with different surveillance schedules and adaptation criteria employed within stories in virtual worlds.

Table 5.4 Stakeholders and their areas of concerns

Stakeholder	Concerns
University President	Campus safety (adjacent areas)
Campus Security	Maximizing campus surveillance with limited resources, pedestrian safety
Pedestrians	Making correct path decisions when traversing campus
Local Police	Juggling routes and timing to respond to campus events/incidents
Paramedics	Responding to campus incidents requiring medical attention

Fig. 5.5 Risk evaluation

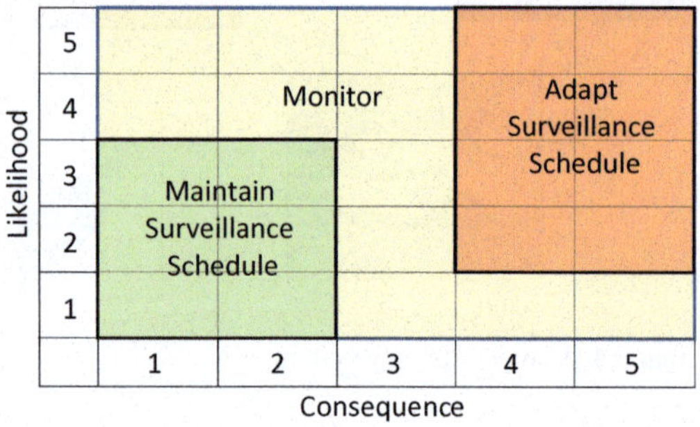

Risk monitoring is an integral part of the campus security system and the coverage it achieves with different surveillance strategies. Figure 5.5 shows the likelihood of an incident and the potential consequence. Three regions are shown in this graph: low incident likelihood-low consequence, high incident likelihood-high consequence, and the regrow between the two. A solution to this problem is to stay with a fixed schedule for low incident likelihood-low consequence and an adaptive surveillance schedule for high incident likelihood-high consequence. The region between the two needs to be monitored to determine if a change in surveillance pattern is warranted.

Equipped with a finite set of monitoring and surveillance resources (e.g., police patrol, cameras, etc.), the campus security system (CSS) assures the security of people on the campus. Figure 5.5 shows the system model represented in SysML. The system model includes requirements diagram, context diagram (scope), graphical use case, behavior diagram, state transition diagram, Block Definition Diagram (BDD), etc. The Unity 3D game engine renders the 3D geospatial, immersive views of the system in the virtual world. These include the campus map with streets, buildings, traffic patterns, and other physical landmarks with dynamic overlays such as tweets and crime statistics by area. The virtual world accepts as inputs the security system, its components (e.g., sensors, processors, actuators, and their properties), and how they integrate and interact with the visualization, animation, and texture generation routines in the virtual world, as well as with scenario background factors such as pedestrian risk levels and target value. The story variables include normal/increased pedestrian traffic, events (campus, adjacent area), campus security system malfunction, behaviors of actors, locale, coverage, and changes in coverage.

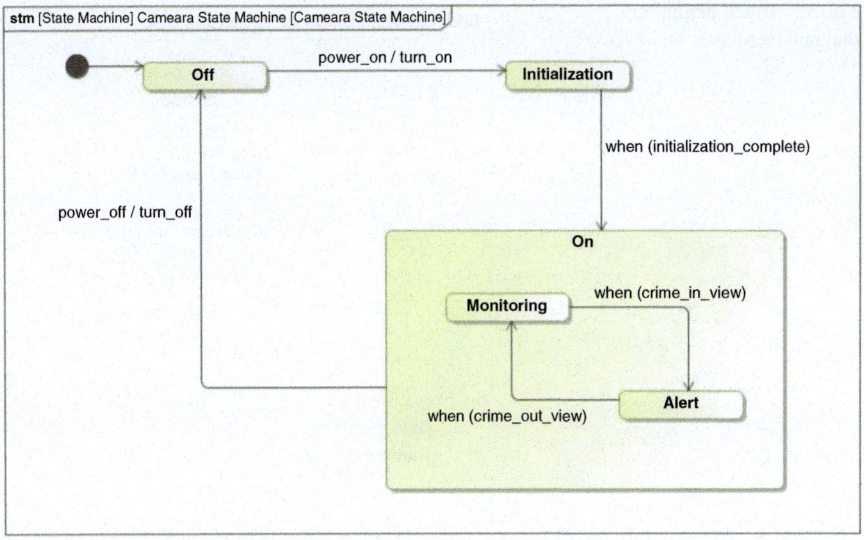

Fig. 5.6 State transition model

A customizable mapping module translates the inputs into visual/multimedia elements (e.g., color, speed, three spatial dimensions) and time. An object interactivity matrix (with the entire scene as one of the objects) represents the possible interactions that can occur during dynamic system operation.

Engineering Model Views A state machine representation depicts a simple behavioral view associated with the camera (an active sensor). This view captures the possible camera states and transition between them (Fig. 5.6).

Figures 5.7 through 5.10 present structural views associated with the CSS. Figure 5.7 shows the block definition diagram (BDD) for the sensors used by the CSS. Figure 5.8 shows the BDD for the key world model objects associated with the campus. These include physical objects, environment, sensors, building, patrol, and crime data. Figure 5.9 shows the BDD for buildings/structures protected by the CSS. These include school buildings, parking structures, and campus restaurants. Figure 5.10 shows the interface definition diagram for the USC campus.

Virtual World View of CSS While the views presented in Figs. 5.7 through 5.10 are appropriate for systems engineers, they are far less meaningful to the broader stakeholder community that comprises mostly nontechnical people. To overcome this problem, we use system stories to communicate system design and CONOPS. In other words, system stories contextualize system design and CONOPS. Stakeholders interact with the stories in virtual worlds and thereby uncover hidden interactions.

In the virtual world, stakeholders (i.e., users) can view and experience the CSS against a geospatial background with appropriate overlays (e.g., animation, annotation). The visual scenes are filtered using sensor types, geographic regions, hotspots,

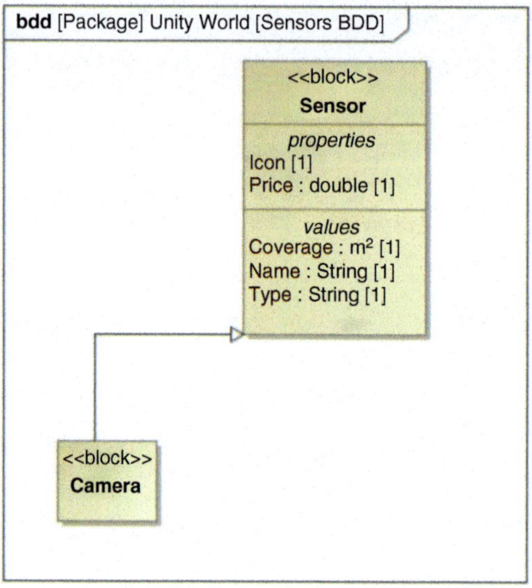

Fig. 5.7 Block definition diagram (sensors)

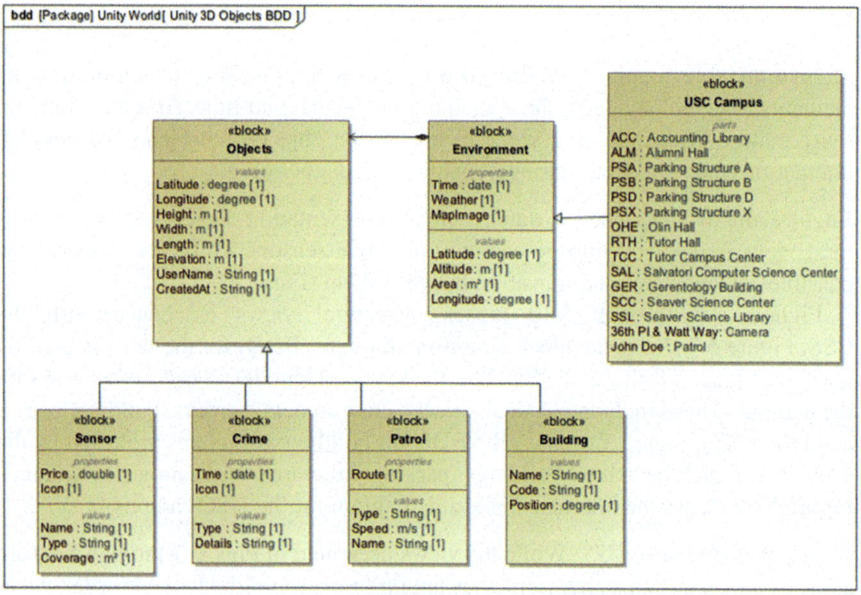

Fig. 5.8 Block definition diagram (3D objects)

and high-traffic areas as filters. It is possible to add a hypothesis generation and testing module, potentially assisted by a virtual AI tutor, to the virtual world's underlying model. The testing module helps users in making choices to test a system, subsystem, or part using playback and specific branching points to test different deployment hypotheses. Another benefit of virtual world interaction is that

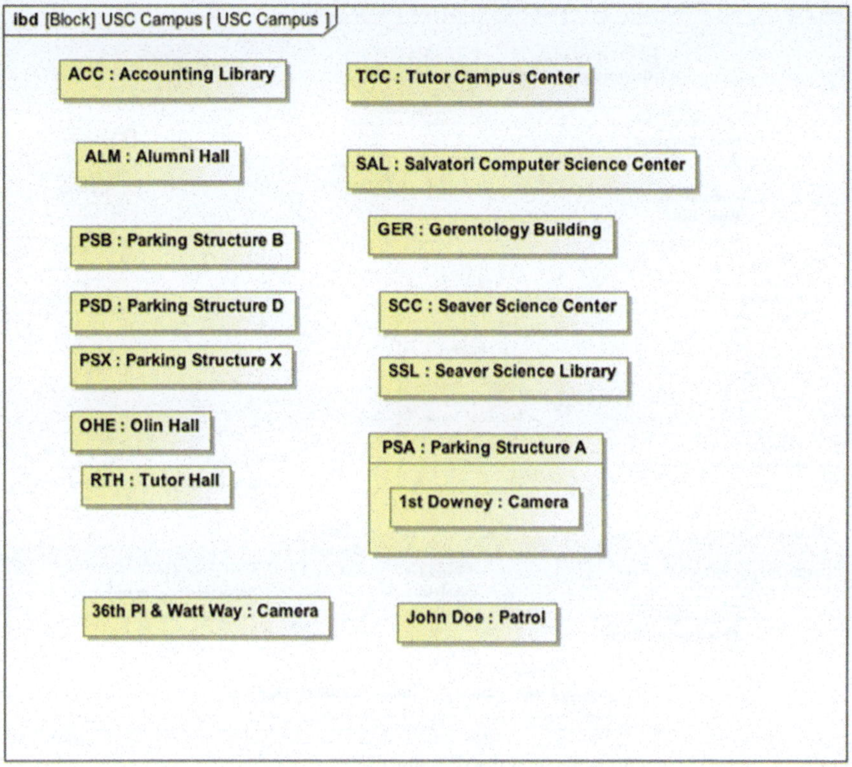

Fig. 5.9 Block definition diagram (USC campus security)

multiple users can view the same/different views of the system and collaborate on a shared desktop.

Stories, set in the virtual world, can follow action-decision point flowcharts with embedded questions for branching (e.g., is risk level at a particular location such as SAL building too high at night?). If so, the story develops along one branch. If not, it develops along another branch. Stories can inform daily schedules and onetime events such as a scheduled narrative about foot patrols or a break-in. "Scripted events" can serve as injects into the story. The narrative engine equipped with a sanity checker would assure story integrity and coherence. For example, changes introduced upstream in the narrative can potentially affect downstream events. The story needs to reflect this effect.

Exemplar interactive narrative diagrams for the policeman, the thief, and the mugger (i.e., actors in a CSS story) are shown in Fig. 5.11 in state machine form. The *policeman* performs foot patrol at a particular location and follows a specific route. In the absence of a crime report, the policeman continues to patrol the assigned route. If a crime is reported, then the policeman rushes to crime scene to talk to victim. This location is most likely where the emergency phone that was used

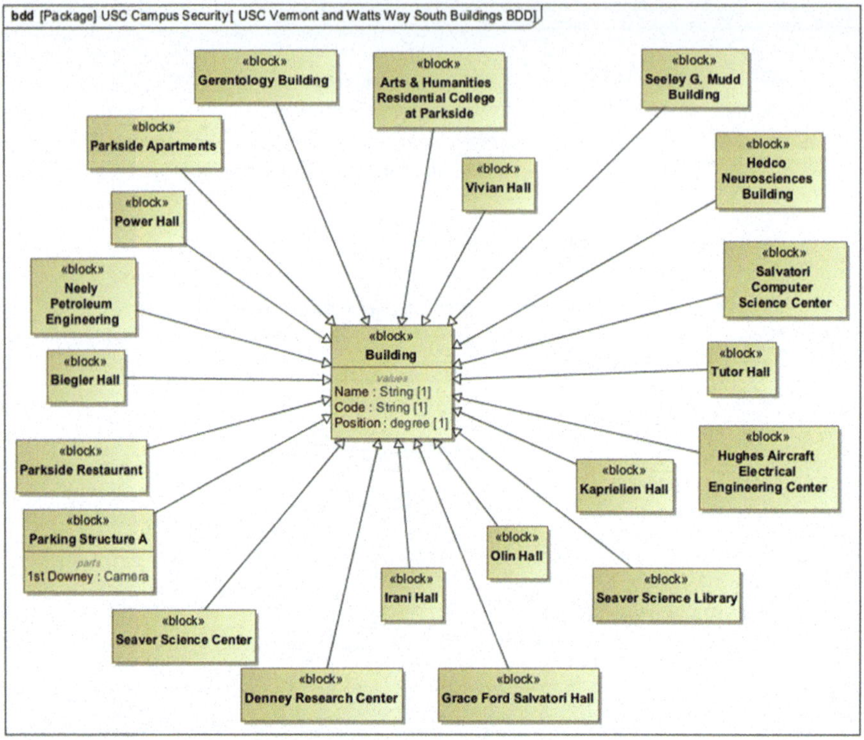

Fig. 5.10 Interface definition diagram (USC campus)

by the student (victim) is located. "Distance to the emergency phone" is the parameter needed to synchronize activities (i.e., the reporting of the crime and the police moving to accost and talk to the student victim). The *student* walks toward the classroom along McClintock Ave at 6:15 pm. In the absence of a mugging report along the route, the student attends the class from 6:30 pm to 9:30 pm. The student performs this routine regularly in the absence of a campus incident (e.g., a mugging). If there is a mugging, the victim (student) walks to an emergency phone and reports the crime. In this case, "distance to the emergency phone" is the synchronizing parameter.

The *mugger* surveils an area for unsuspecting potential victims. If a student is located in the vicinity with no police support or foot traffic around, then the mugger swiftly moves in the direction of the student and locates a concealed spot from which to mug the student. The "distance between the mugger and the student" is the synchronization parameter. In the absence of a potential victim showing up in the vicinity, the mugger continues to surveil the area. Thus, the policeman, the student, and the mugger perform their respective routines (e.g., walking, watching, lying in wait) in the absence of a campus incident. As scheduled and unscheduled events occur over time, the story comes to life.

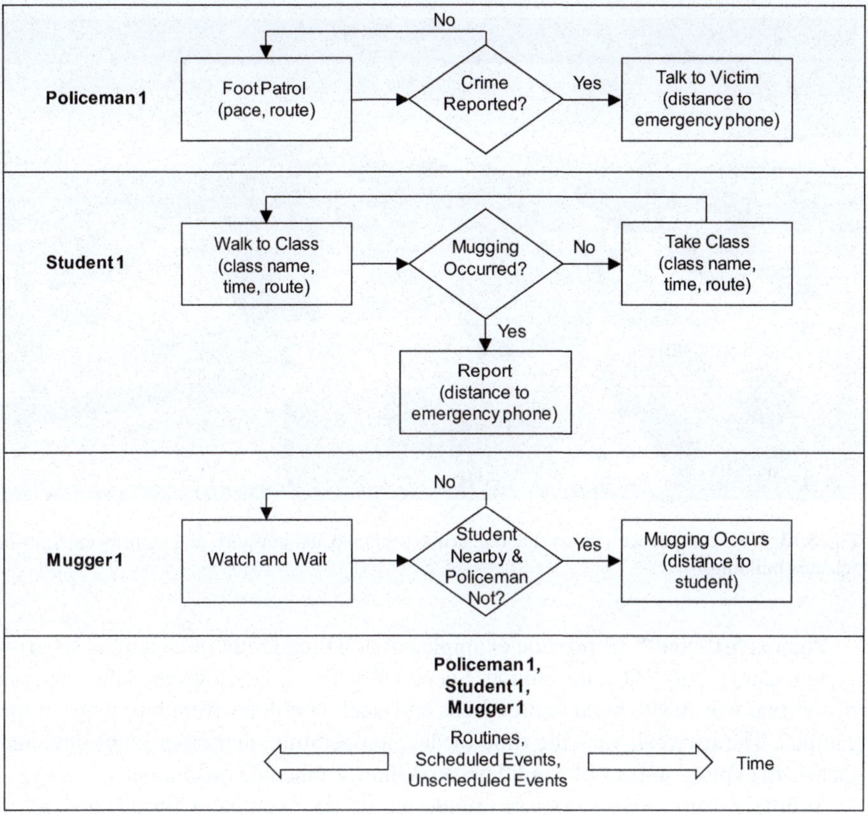

Fig. 5.11 State machine description of interactive narrative

Fig. 5.12 USC campus with local twitter feeds and satellite uplinks

Fig. 5.13 USC campus with dawn shading, crime heat map and hotspots, and security cameras in selected buildings

Figures 5.12 and 5.13 provide examples of a 3D rendering of a campus security system using Unity 3D game engine. Specifically, Fig. 5.12 shows the USC campus in a virtual world with local twitter feeds and satellite uplinks from buildings on the campus. The user can vary the time of day, temperature, and other environmental factors to explore impact of candidate surveillance schedule on campus coverage.

Similarly, Fig. 5.13 shows a 3D rendering of USC campus with dawn shading and crime heat map and hotspots, along with the location of security cameras in buildings on the campus. The user can experience this view under various conditions (e.g., time of day, visibility, temperature, other factors) to uncover surveillance gaps in candidate surveillance schedules.

5.9 Story Development

The story generation components comprise a narrative editor, underlying story formalism, skeletal story authoring (action event builder), and dynamic interactivity matrix. The story vignette in this illustrative example revolves around the CSS example discussed earlier with a student walking on the campus to the engineering building to take a class.

Story Vignette It is late evening and pedestrian traffic has dropped off on USC campus. A student is walking to his class. All appears to be quiet with no security police in the immediate vicinity. A mugger is lying in wait behind the brush and foliage adjacent to the walkway. Taking advantage of the sparse pedestrian traffic, the mugger swiftly and surreptitiously moves in the direction of the student from behind.

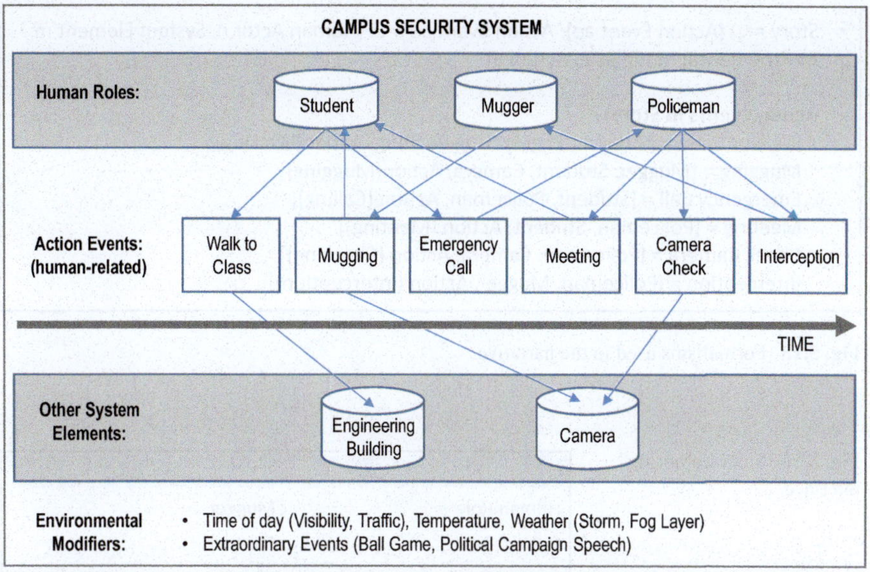

Fig. 5.14 Narrative editor

Moments later, he mugs the student. The mugging is recorded by a camera positioned on the southwest corner of the engineering building. The camera is part of the campus security system. The distraught victim contacts the campus security using an emergency line. The campus security dispatches a police officer on patrol to assist the victim and then pursue the attacker. The officer briefly meets the mugging victim at the engineering building, makes sure that the victim is not in need of medical attention and has recovered sufficiently from the shock of the mugging, and reads the appropriate camera log. He then takes off to apprehend the attacker and succeeds in apprehending and arresting him a short distance away.

Narrative Editor The narrative editor allows the user to define a high-level view of the narrative within the operational context defined by the environmental factors. The CSS comprises human roles that participate in the story vignette, as well as other system elements within the CSS. A partially ordered, temporal sequence of action events defines the story vignette. The narrative editor allows the story author to edit a subset of variables before bringing the story to life in the virtual world (Fig. 5.14). This story vignette, shown in Fig. 5.14, has three actors (i.e., student victim, mugger, policeman) and six action events (i.e., student walking to class, mugging, emergency call, meeting, camera check, interception) that occur over a timeline. Other system elements include buildings that house cameras and the cameras themselves. The environmental modifiers are time of day that defines visibility and traffic, temperature (clothing), weather (rain, fog), and extraordinary events (e.g., ball game, political campaign speech). The user can edit the story in terms of actors, action events and their times of occurrence, other system elements, and environmental modifiers.

- Story = ∪ {Action Event ae}; Action Event ae = ∪ {Human Actor n, System Element m, Environmental Modifier e, Action a}
- **Action Events in Story:**
  Walk to Class = [Student, Engineering Building, Action(Walking)]
  Mugging = [Mugger, Student, Camera, Action(Mugging)]
  Emergency Call = [Student, Policeman, Action (Calling)]
  Meeting = [Policeman, Student, Action (Meeting)]
  Check Camera = [Policeman, Camera, Action (Checking)]
  Interception = [Policeman, Mugger, Action (Interception)]

Fig. 5.15 Formalisms used in the narrative

Fig. 5.16 Story authoring interface

Action Event:	Mugging
Human Role	Mugger
Human Role	Student
System Element	Camera
Action	Mugging (Mugger, Student, Camera)
Time	8:45PM
Place	@Engineering Building, SW corner

Formalisms Used Figure 5.15 presents the formalism used to represent the narrative.

Skeletal Story Authoring Drilling down from the narrative editor, enables story authoring. Typical story elements include action event, human role(s), system elements, action, time, and place. Figure 5.16 presents the interface for story authoring in terms of building action events.

Dynamic Interactivity Matrix The dynamic interactivity matrix (DIM) captures the potential interactions among story elements (i.e., actors, environment, other system entities, and action events). A story that is interactively executed in a virtual world enables "what-if" exploration of system behavior to uncover hidden interactions among system elements and between the system and the environment and use that knowledge to update the DIM and the system model. Thus, the DIM starts with known interactions in the model. The interactions uncovered through "what-if" story exploration update the model. In this sense, the DIM's characterization is having an adaptive capability. Figure 5.17 shows a sample DIM. In this figure, the column affects the row in characterizing action events.

Column affects Row	Student	Mugger	Policeman
Student	N/A	Mug	Meet
Mugger	N/A	N/A	Intercept
Policeman	Call	N/A	N/A
Camera	N/A	Appear on	Check
Engineering Building	Walk to	N/A	N/A

Fig. 5.17 Example interactivity matrix

A DIM can be drilled down from the perspective of an engineer or a nontechnical stakeholder. The engineer perspective consists of state machines, block definition diagrams (BDDs), and interface definition diagrams (IDDs). The nontechnical stakeholder perspective consists of agents, locations, times, and behaviors.

5.10 Summary

Not all stakeholders understand systems engineering models. Even fewer understand systems engineering notation. As a result, non-technical stakeholders are unable to fully participate and contribute to early design and upfront engineering where these models and notations are used. This deficiency invariably leads to extraneous design iterations and rework. In this chapter, I showed that by transforming system models into system stories, all stakeholders can be meaningfully engaged. Specifically, the different stakeholders can interactively execute their own stories in virtual worlds and thereby increase their understanding and contribution to upfront engineering. I called this model-based interactive storytelling (MBIS).

MBIS begins with authoring partially scripted stories about complex systems. Stakeholders can then interact with those partially scripted stories to explore complex system behaviors and change propagation paths. In the process, they can uncover system hotspots and unintended consequences. The central idea behind MBIS is to allow stakeholders to "experience" the behavior of complex systems using stakeholder-specific lenses and with facilities to rewind and replay stories with specific "injects." During replay, they can pause and resume at key points, and conditionally branch and loop to explore system behavior further. These capabilities serve to not only engage stakeholders but allow them to create a mental picture of system behavior in different contexts. MBIS impacts complex systems engineering in a number of ways. These include uncovering hidden interactions and dependencies within the system, ensuring that all stakeholders contribute to collaborative design especially in upfront systems engineering, and exploring alternate futures with different technologies, assumptions, initial conditions, and CONOPS. Finally, MBIS can be combined with complementary complex systems engineering approaches such as graph-based/matrix-based methods, agent-based modeling, and set-based design to create a comprehensive capability for system modeling, computer-aided critiquing, system analysis, system behavior exploration and visualization, and system redesign and management.

References

1. Greenhalgh T, Hurwitz B. Why study narrative? BMJ. 1999;318(7175):48–50.
2. Bowman DE, Ortega RE, Hamrick ML, Spiegel JR, Kohn TR. Refining search queries by the suggestion of correlated terms from prior searches. Google Patents; 1999.
3. Perlin K, Goldberg A. Improv: a system for scripting interactive actors in virtual worlds. Proceedings of the 23rd annual conference on Computer graphics and interactive techniques; 1996: ACM.
4. Bartle RA. Designing virtual worlds: New Riders, Indianapolis, Indiana, 2004.
5. Madni AM, Andrecut M. Efficient heuristic approach to the weapon-target assignment problem. J Aerosp Comput Inf Commun. 2009;6(6):405–14.
6. Madni AM. Expanding stakeholder participation in upfront system engineering through storytelling in virtual worlds. Syst Eng. 2014;18(1):16–27.
7. Madni AM, Nance M, Richey M, Hubbard W, Hanneman L. Toward an experiential design language: augmenting model-based systems engineering with technical storytelling in virtual worlds. Procedia Comput Sci. 2014;28:848–56.
8. Madni AM, Richey M, Ordoukhanian E, Venkatesh J, Zender F, Chang K, et al. Exploiting storytelling in collaborative systems engineering: towards a smart experiential dashboard. Conference on Systems Engineering Research Huntsville, Alabama 2016.
9. Madni AM. Model, stories, immersive experience: system engineering in 21st Century, INCOSE Regional Mini-Conference, Keynote talk. 2016.
10. Orellana D, Madni A. Extending model based systems engineering for human machine interaction analysis and fault tolerant design. Infotech@Aerospace 2012; 2012/06/19: American Institute of Aeronautics and Astronautics (AIAA); 2012.
11. Madni AM, Madni CC. Rapid scenario building and optimization for simulation-based exercises. 2009 IEEE Conference on Technologies for Homeland Security; 2009/05: Institute of Electrical & Electronics Engineers (IEEE); 2009.
12. Graesser AC, Lang KL, Roberts RM. Question answering in the context of stories. J Exp Psychol Gen. 1991;120(3):254.
13. Clarke A, Jane Hanson E, Ross H. Seeing the person behind the patient: enhancing the care of older people using a biographical approach. J Clin Nurs. 2003;12(5):697–706.
14. Riedl MO, Sugandh N. Story planning with vignettes: Toward overcoming the content production bottleneck. Joint International Conference on Interactive Digital Storytelling; 2008: Springer.
15. Riedl MO, Stern A. Believable Agents and Intelligent Story Adaptation for Interactive Storytelling. Technologies for Interactive Digital Storytelling and Entertainment: Springer Science + Business Media; 2006. p. 1–12.
16. Riedl MO, Young RM. From linear story generation to branching story graphs. IEEE Comput Grap Appl. 2006;26(3):23–31.
17. Thorne JP. George Lakoff , Mark Johnson, Metaphors we live by. Chicago and London: The University of Chicago Press, 1980. Pp. xiii + 242. Dwight Bolinger, Language the loaded weapon: the use & abuse of language today. London and New York: Longman, 1980. Pp. ix + 214. J Ling. 1983;19(01):245.
18. Madni AM, Spraragen M, Madni CC. Exploring and assessing complex systems' behavior through model-driven storytelling. 2014 IEEE International Conference on Systems, Man, and Cybernetics (SMC); 2014/10: Institute of Electrical & Electronics Engineers (IEEE); 2014.

19. Madni AM HUMANE: a knowledge-based simulation environment for human-machine function allocation. Proceedings of the IEEE 1988 National Aerospace and Electronics Conference: Institute of Electrical & Electronics Engineers (IEEE); 1988.
20. Carlson LE, Sullivan JF. Hands-on engineering: learning by doing in the integrated teaching and learning program. Int J Eng Educ. 1999;15(1):20–31.
21. Nach H, Lejeune A. Coping with information technology challenges to identity: a theoretical framework. Comput Hum Behav. 2010;26(4):618–29.

Chapter 6
Novel Options Generation

> *"A world which sees art and engineering as divided is not seeing the world as a whole."*
>
> – Professor Sir Edmund Happold, structural engineer

6.1 Background

Option generation, and especially generation of novel options, is a critical capability that is needed during up front engineering and especially during systems architecting. Systems architecting is an integrative, decision-rich activity that requires the generation and evaluation of novel options to exploit technological advances and assure compliance with changes in programmatic and institutional constraints. From a decision making perspective, option generation circumscribes the set of potential decisions/actions available to the decision maker. It makes the decision maker (e.g., system architect) define the decision parameters of the problem in precise terms. It sets upper and lower bounds on the quality of the decision maker's choices. Surprisingly, option generation is not given the attention it deserves because many decision-theoretic models assume that the decision maker has a predefined set of options [1]. With this mindset, the problem becomes one of representing available options in a formal decision support framework [2, 3]. Even when the assumption of having a predefined set of options at the start was removed in subsequent methods [4], option generation remains a difficult task because it requires substantial knowledge of the problem domain (e.g., air traffic control, healthcare) and understanding of the decision context. Reducing the effort required to extract, process, and represent knowledge in usable form is a key challenge in option generation today.

There are other challenges as well. First, relevant information stored in a database needs to be recognized as relevant to option generation and then retrieved, translated, and formatted for presentation in a useful form. Second, human memory tends to be imperfect and biased toward the contextual specifics of the time that information was collected, assimilated, encoded, and stored. Consequently, efficient information recall requires presentation of the associated context, i.e., context at the time the information was encoded. Third, human judgment tends to be fallible. Specifically, research in decision analysis suggests that humans tend to be relatively insensitive to portions of the problem that may have been inadvertently left out,

A.M. Madni, *Transdisciplinary Systems Engineering*,
DOI 10.1007/978-3-319-62184-5_6

thereby tacitly but erroneously assuming "completeness" of the option set [5]. In other words, upon viewing information in a certain form and format, humans tend to be limited in their ability to envision and supply what's missing. In fact, the ability of humans to interpolate across gaps in problem representation can often be severely limited without explicit amplification/augmentation of cognitive processes. Finally, the formal structure within which option generation fits does not suggest the content of the options. Not surprisingly, both humans and smart decision aids continue to experience difficulty in generating novel options.

6.2 Novel Option Generation

Novel options, by definition, are options without precedent, or those with unique characteristics. Developing novel options can be facilitated by having a conceptual framework within which multiple option generation aiding techniques can be employed [6]. However, developing such a conceptual framework poses several challenges. First, decision scientists tend to view novel option generation as a creative process outside the purview of normative decision making techniques [7]. Second, having individuals simply enumerate options tends to be inadequate because people tend to fixate on the problem representation provided to them [8]. In fact, psychological studies have shown that even in the most benign circumstances, humans tend to focus on a narrow range of alternatives. Third, the experiential knowledge base of an individual is hard to access when generating options. Fourth, while formal methods and models provide a convenient means for representing options, they neither suggest options nor provide the means for creating novel ones.

In light of the foregoing, option generation during systems architecting is a challenge that cannot be ignored. Having a systematic process for option generation helps ensure completeness of the option space which is essential for making effective architectural (and design) trade-offs. It enforces a certain structure that causes decision makers to precisely define problem parameters. It promotes understanding and facilitates communication of the decision problem [9, 10]. While option generation comes into play throughout the system life cycle [11–13], its impact during early design, and upfront engineering in general, is especially impactful. In particular, option generation plays a key role in concept of operations (CONOPS) development, interface design, human-system integration, integration architecture development, legacy integration, and testing. Table 6.1 presents a representative set of parameters that characterize the decision maker's option space.

Real-world constraints on option generation are both human related and context related. *Human-related constraints* tend to be mostly psychological in nature. They include the inherent difficulty of humans in articulating and circumscribing a decision situation, human cognitive limitations, limited ability to organize and manage generated options, limited ability to make best use of past experiences, and limiting assumptions/beliefs about a decision situation. *Context-related constraints*, which

Table 6.1 The option space

• Objectives and Goals
• Needs Characterization
• Regulations and Policies (existing, expected)
• Technologies (readiness levels, expected maturity rates)
• Technical Constraints (legacy; test and evaluation)
• Programmatic Constraints (schedule, funding, space, location, acceptance criteria)
• Operational Constraints (environment, terrain, weather, rules of engagement, human limitations)

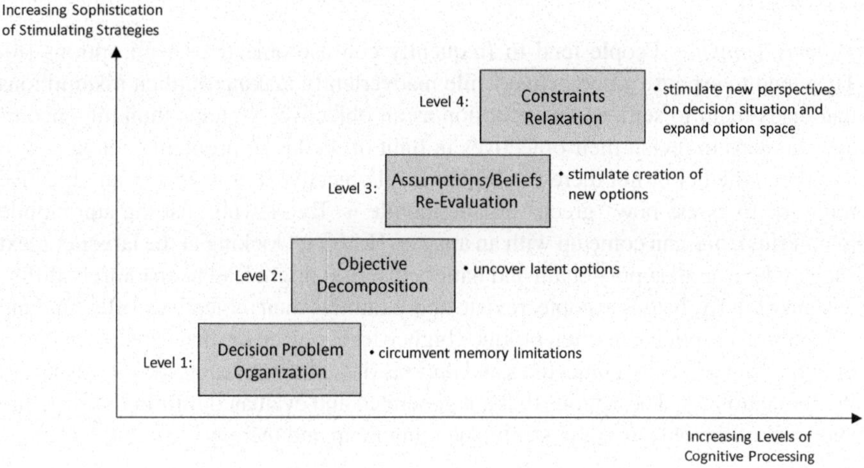

Fig. 6.1 Framework for enhancing novel option generation

are primarily environment related, include organizational rules and hierarchies; institutional biases and orthodoxies; structure and content of social and cultural interactions; and, occasionally, geographic distance. It is against this backdrop that the multilayered framework for enhancing option generation was developed (Fig. 6.1).

The unified framework, depicted in Fig. 6.1, is amenable to computer-based aiding. It employs four levels of cognitive processing to stimulate option generation. Level 1 is concerned with *organizing* the variables associated with a decision situation and modeling the decision situation. By having an explicit representation of the decision situation, human memory limitations can be circumvented. Level 2 is concerned with *decomposing* the objective as a means to uncover latent options buried within the formal representation of the problem structure. Level 3 is concerned with *questioning* and evaluating assumptions, constraints, beliefs about the prevailing context/environment, and the decision situation, with a view to further stimulating creative option generation. Level 4 is concerned with *relaxing* constraints on the option space. Examples of constraints that can be relaxed are resource limitations, schedule, organizational hierarchy, human skill level, funding limitations, and business policies. Systematically relaxing these constraints and then examining the option space can stimulate new perspectives on decision situations that arise during systems architecting and early design.

6.3 Principles for Stimulating Novel Option Generation

Madni et al. [8, 14] presented a set of domain-independent principles for stimulating option generation. An extended set of these principles are presented and discussed next.

Systems Thinking People tend to frequently confuse objectives with options [8]. They tend to espouse an objective while inadvertently making implicit assumptions that leads them to settling for an option as an objective. Systems thinking encourages humans to revisit their objectives in light of the larger problem context and to ascertain whether or not there is a higher-level objective. For example, an objective might be to assess how "green" an automobile is. By viewing just the automobile (e.g., Prius), one can come up with an answer. However, looking at the larger context such as the overall supply chain and battery disposal might lead to an entirely different answer. By having people revisit and examine their objectives after making assumptions explicit can often produce higher-level objectives that capture their true intent. Cycling between objectives and options (i.e., alternatives) is the key to enriching the option set. The options that are generated and evaluated inform the structure and level of the objectives, possibly changing them and thereby closing the cycle.

Checklists Checklists, created in early development, consist of objectives and a set of default options associated with that objective. By presenting a checklist of frequently used or generic options (including the "status quo" option), human decision makers can be offloaded from having to recall these from memory. It is well known that in some situations humans can fail to recall options for simple and repeatedly performed tasks [15]. Since human memory recall is fallible, humans tend to employ a variety of internal memory aids (e.g., stories, visual memory joggers) and external memory aids (e.g., memos, lists). However, despite awareness of internal memory aids, it appears their use is infrequent [16]. Thus, the value proposition of internal memory aids is questionable at best. For this reason, it makes sense to rely on external memory aids such as checklists. Checklists also ensure that in the pursuit of novel options, decision makers don't ignore obvious options. A checklist can be expected to evolve over time, growing in number through the addition of new and specialized options (i.e., options specialized for different decision situations). An example of effective use of a checklist is in sensor selection for a mission with specific requirements. Associated with the mission objective are a set of default sensors that can detect objects at various ranges. When the user specifies a mission objective and calls up sensors for that mission, the full pre-stored list of sensors shows up for that mission. Then, the user starts to review the constraints on sensor selection (e.g., sensor needs to be integrated into an air vehicle that has limitations on payload weight and size; the probability of detection needs to be at the individual person level). Review of these constraints helps prune the set of sensors to the viable few. Beyond that the user may conduct trade-offs between altitude, weight, and probability of detection, Pd. The results of such trade-offs can further prune and help prioritize the sensors in the checklist for the specific mission objective.

Decision Situation Modeling (DSM) This technique is concerned with "framing" the design problem [17]. Explicitly modeling a decision situation in terms of options, uncertainties, consequences, and expected values and selectively focusing on each characteristic can expand the option set. By having the decision maker identify anticipated consequences, enumerate and assess uncertainties, identify riskiest, then attempt to generate even riskier ones, and then generate expected values of decision/action alternatives, the option space can be expanded [18, 19]. For example, let us say that we have specified a sensor system for development. Within the specified parameters, we come up with design alternatives. We ask ourselves if there are any more design alternatives. We then investigate how the specification was developed. This investigation can shed new light on the specification (e.g., the problem was too tightly specified, the specification was too narrow or had an error) and potentially lead to the creation of new design alternatives.

Subjective Probability Analysis Making the knowledge associated with an option explicit improves probability assessments (by reducing uncertainty) and stimulates broad thinking about a decision problem [20]. The quality of probability assessments is improved by considering both supporting and contradicting evidence. Grouping evidence that supports/contradicts the effectiveness of an option stimulates associative mental activity that can potentially stimulate the generation of new options. In the same vein, grouping evidence that supports a common set of options can potentially help define the option space and encourage thinking about new options. Similarly, grouping evidence that contradicts a common set of options can potentially reveal inadmissible portions of the option space. The intent of this strategy is to overcome "failure of imagination" by looking at the decomposition from the point of view of probability distributions, rather than as a point problem. The emphasis is on making the underlying assumptions explicit and looking for distributions of possible outcomes rather than researching only the best case scenarios. By examining the assumptions underlying an alternative, and systematically varying the assumptions, it is possible to assess what the alternative is likely to achieve (i.e., how robust the alternative is). For example, let us say that we are designing a robotic system for a particular mission. Let us say that we come up with a design and not take into account the reliability of the different components and the probability of failure in the design. When we finally do, we can end up with different design alternatives that turn out to be more robust or resilient than the original option.

Problem "Back Translation" This technique is motivated by the recognition of an inherent problem that exists when people work in groups. When individuals communicate descriptions of problem situations to others, certain characteristics of the situation get highlighted or altered in the transmission [21]. When the situational information is retransmitted by the recipient to the sender, a new perspective on the original problem can potentially emerge. Thus, options along with option generation parameters (e.g., objectives, constraints) can be transmitted to others for redefinition. "Back translation" introduces novel perspectives on the problem and helps reveal ambiguous elements. These can lead to new option generation opportunities. The capability for translation and communication of decision situations to foster option generation is enabled through network connectivity and intelligent networking architectures. For example, let us consider the design activity. The design activity, simply stated, is the translation of a specification into a design. Now consider translating a specification into design and forwarding the design to a member of the collaborative design team. This individual reviews the design, back-translates it into what the specification might have been, and sends that specification to the sender. Upon comparing the original specification with the "back-translated" specification, more often than not a discrepancy will be discovered. This discrepancy can offer insights into creating new design alternatives.

Opponent or Competitor Analysis Individuals typically perceive themselves as behaving on the basis of a different set of principles than others placed in the same situation. Specifically, people tend to view themselves as responding more to circumstantial factors than others placed in the same situation would. This analysis begins with conducting decision situation modeling (DSM) for both own situation and that of the opponent or competitor. This is followed by determining the set of environmental factors that might drive the behavior of the opponent or competitor. Then, implications of the competitor's/opponent's attributes such as constraints and resources are explored relative to the development of one's own option set. This process can potentially enrich one's option set. For example, consider the competitors to one's new product. The competitors are faced with a set of questions (e.g., who to partner with, what standards to adopt immediately, what regulations to comply with, etc.). By systematically performing "what-if" analysis for each such variable after prioritizing them can enrich one's option set.

Temporal Analysis When things don't work, it always appears that the outcome was foreseeable. Imagine 2 years out, a system that is operating normally fails. How will we have looked at the design that we have created now? Consider NASA's Mars mission and the Seven Minutes of Terror. The Curiosity rover was poised to hit the top of the Martian atmosphere at 13,000 mph. If all went according to plan, the rover was to be slowly lowered by cables into a massive crater in the final few seconds. NASA's plans for Curiosity called for a series of braking maneuvers, similar to those employed by the Space Shuttle, and a supersonic parachute to slow it down. Then the heat shield for the fiery descent was to be discarded. Then the rover was to be lowered by cables from a hovering rocket-powered "sky crane." At touchdown, the cords were to be cut and the sky crane was supposed to crash at a distance. Everything had to go like clockwork in this mission which, by any measure, would be considered highly complex. Fortunately, the mission did go like clockwork. But what if the mission didn't work as expected? What would the system engineers have looked at? Would the current design be still what they would go with? Should they have considered an alternative design? It is worth noting that they were working to an outcome that they wanted to achieve from the alternatives that they had chosen. They would need to shift their perspective from a point outcome to a distribution of outcomes. So, they would need to "move forward" in time and imagine certain unintended outcomes and ask themselves if the design still works. This is what temporal analysis is all about. It is about shifting the psychological frame of reference around the design problem (e.g., specification, outcome, translation of specification into design, temporal perspective and unintended outcomes, problem evaluation method).

In sum, the temporal perspective from which a situation is viewed influences one's judgments about the situation. For example, events seem more probable in hindsight than in foresight. In foresight, many pathways appear to lead to an objective, while in hindsight only one pathway appears probable. Similarly, anticipated gains can outweigh anticipated losses in a future time horizon rather than in the immediate time horizon. Specifically, temporal analysis consists of (i) modifying (adding/deleting) option generation parameter sets according to how a problem situation would appear if set in alternative time frames, (ii) exploring the consequences of candidate options in both near-time and far-time horizons, (iii) generating options for future time frame, and (iv) then applying the attributes of the future option set to the present situation to enhance the option set for the present situation.

Causal Modeling Humans tend to rely on mental organization to track and manage complexity. This is problematic in that humans tend to be linear thinkers, whereas system management requires complexity systems thinking, an inherently nonlinear activity. Fortunately, humans are good at articulating causal chains. Thus, causal modeling (e.g., fault tree analysis, event tree analysis, system dynamics model) can be employed to enhance understanding of how a particular system state can or should come about. The psychological principles that inform causal modeling are that (a) humans, in fact, can adequately map a set of primitive parameters of a problem (e.g., an objective) to a set of conditions that can be realized in terms of decisions/actions or events, but (b) analysts perceive problem analyses to be more complete than they actually are. The implication of causal modeling is that objectives need to be reevaluated and redefined in light of their degree of fulfillment by candidate options. For each option, the steps involved in achieving objectives need to be traced. Furthermore, asking analysts directly to imagine how and where problem analyses could be incomplete can potentially stimulate new perspectives that result in generating new options. In sum, human capabilities should be exploited to articulate causal chains while circumventing human limitations (e.g., inability to recognize how probabilities fall off, inability to make judgments about entities that change exponentially).

Debiasing Institutional Inhibition Decisions are made in the context of institutions that include social norms and culture. The influence of organizational constraints in design can be powerful. Thus, the problem context often poses unacknowledged/inadvertent constraints on option generation (e.g., respect for organizational hierarchies/echelons, restrictions to doctrinal objectives and options), thereby inhibiting option generation and preventing realization of the actual size and content of the option space. For example, a system architect/engineer working in an organization with formal policies and procedures might feel inhibited to express a good idea that might appear to violate an organizational policy. In such cases, time and space need to be set aside where individuals can freely express these ideas without fear of reprimand. Therefore, institutional constraints need to be separated from real situation-related (i.e., contextual) constraints. "Unthinkable" objectives and options should be allowed into the discourse about the problem domain with a view to uncovering options that may, in fact, turn out to be acceptable and effective. "As-if" techniques should be employed to enable option generators (i.e., decision makers) to assume higher or lower levels of authority than they actually have.

Analogical Reasoning Reasoning by analogy is viewed as a core aspect of human cognition [4]. It is important for learning and classification and for thought and explanation. Analogical reasoning is involved in key scientific discoveries, and, in fact, many argue that analogies are central to creative thinking. Holyoak and Thagard [22] introduced the concept of "mental leaps" to characterize analogies. They argued that the act of forming an analogy requires seeing one thing as if it were another. In so doing, the individual makes a kind of "mental leap" between domains. In fact, mental leaps are evident in well-known scientific analogies. A famous example of analogical reasoning is Archimedes' insight that water displaced by an object can be used to quantify the mass of different objects. Analogical reasoning looks beyond traditional engineering boundaries into other disciplines such as biology, sociology, and culture to discover examples, paradigms, and metaphors that may well offer a solution to the problem at hand. In this way, analogical reasoning can be exploited to uncover novel options [23, 24]. For example, the human immune system offers an apt analogy for building a cybersecurity infrastructure.

Sensitivity Analysis Sensitivity analysis, a natural outgrowth of decision analysis, tests the robustness of a problem analysis to variations in its input parameters. Sensitivity analysis highlights alterations in assumptions about factors such as objectives, constraints, states of the world, facts, user beliefs, and values. Alterations in problem assumptions facilitate the opening of "mental locks" that inhibit creative option generation. To this end, the specific strategies that facilitate this process include systematically altering problem features and option generation parameters, asking the user to assume that some (or all) constraints exist or changing the probability that a particular constraint exists, encouraging users to assume different risk levels, and asking users to assume nonexistence of certain relevant facts (e.g., intelligence reports). These perturbations can illuminate new options that can be added to the option set [7]. For example, let us say a system design is based on a mature technology (i.e., Technology Readiness Level 7 or higher). The design solution based on this technology provides a certain level of effectiveness comparable to that offered by the competition. Now let us say that a promising technology at TRL 3 is substituted for the mature technology. The result of the sensitivity analysis in this case clearly shows the value proposition of this technology, but there is the risk that the technology is immature. In this case, the system architect/engineer may incorporate a real option [25, 26] relative to this immature technology by investing modestly in the infrastructure to make such substitution possible at a later date. If the technology matures in time, the architect can exercise the real option. If the technology does not mature in time, the architect can pass on exercising the option. This is an example of how sensitivity analysis can inform and shape strategic decision making.

Option Management One of the problems that occurs in complex organizations (e.g., an engineering organization) is that, all too frequently, the organization pursues a course of action without making explicit why other paths (i.e., courses of action) were not taken. For example, it is imperative that if Option B was discarded in favor of Option A, then that fact and the supporting rationale need to be documented. This capability falls under the rubric of option management. As important, every option has a life as we look down the road, and every decision alternative is a good idea up to a point. In other words, we can't go by "goodness of alternative." Rather, we need to work with probabilities. It is important to realize that at some point, every good decision alternative will look bad or fail. So, how do we protect against this reality? We need to look far enough and understand how long a particular option remains viable before the need to reevaluate it arises. Option management conceptualizes the option generation process in terms of an evolutionary "life cycle" that spans: option generation, option representation (i.e., option expression in a language consistent with the decision problem), option evaluation (i.e., assessment of option quality), option implementation (i.e., option reduction for communication, recording, and alteration), and option control (i.e., option maintenance, learning from experience, and process iteration).

6.4 Option Generation Aid

An option generation aid must have some referent to the architectural trade-offs made and decision making history of the performing organization. Options generated for similar past situations can be recalled for consideration. Options eliminated from the option set are retained for future references for either the same or different problem situations.

The key principles for stimulating novel option generation are employed in the option generation aid shown in Fig. 6.2. The option generation aid is intended to assist decision makers (e.g., system architects, program managers) in rapidly retrieving available options and generating novel ones during systems architecting and design. The system concept of the aid is based on the conceptual framework (Fig. 6.1) and principles for stimulating novel option generation.

Users of the aid are system architects, system designers, system engineers, and program managers. The *user interface* of the aid employs templates, menus, and graphs. Templates are used by the user to fill in the particulars of an option, menus are used to choose an option or attribute, and graphs are used to depict relationships among objectives and options and between options. The *option generation process*

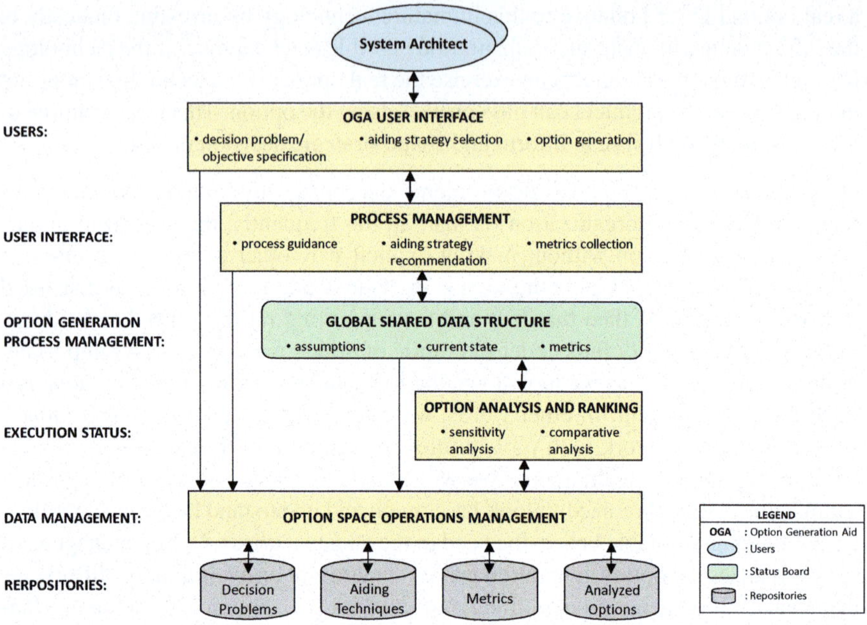

Fig. 6.2 System concept and layered architecture of option generation aid (OGA)

management layer supports invocation of interactive option generation aiding technique, process monitoring, and metrics collection. The option stimulation techniques range from decision problem organization to objective decomposition to assumptions/beliefs reevaluating to constraint relaxation. The constraints that are systematically relaxed include temporal, financial, structural, physical, and environment constraints. The *shared global data structure* ("blackboard") maintains the execution status of the option generation process. It maintains status of variables such as current objective, known options associated with the objective, aiding strategies used, assumptions and constraints relaxed, new options generated, and metrics collected. The *option analysis and ranking* layer supports sensitivity analysis and comparison of options and produces a prioritized option set for the prevailing objective. The *option space operations management* layer supports the various operations performed on options including definition, indexing, storage, linking (to objectives and other options), and retrieval (Table 6.2). The operations performed on the option space, informed by past research of Madni et al. [8] and Altshuller [4], are presented in Table 6.2.

The *repository* layer comprises decision problems and objectives, pre-stored options, various restrictions that generated options needed to satisfy (e.g., constraints associated with technology (e.g., TRL), legacy components, interfaces, schedule, and cost), option generation process model (i.e., stages, entry/exit criteria, resource required), option generation aiding techniques (e.g., systems thinking, checklists, decision situation, modeling, causal modeling), metrics collection (e.g., process metrics, outcome metrics), and online catalog of analysis results.

Table 6.2 Operations performed on the option space

• *Option definition* – defining an option in domain-independent or domain-specific terms
• *Option generation* – the selection of existing options and creation of new ones
• *Option indexing/metadata tagging* – associating attributes with options to facilitate their access
• *Option encoding* – associating contextual attributes at the time option was created
• *Option storage* – persistent storage of options in a repository
• *Option linking* – interrelating and cataloging options that share common attributes
• *Option tailoring* – customizing an option for a particular problem
• *Option retrieval* – fetching an option from an option repository based on key attributes
• *Option space management* – tracking options and monitoring option space throughout option generation life cycle
• *Option presentation* – presentation of an option in textual/visual form to the decision maker
• *Option sensitivity analysis* – varying specific inputs to an analysis and observing effects on the recommended option
• *Option ranking* – prioritization of options based on user-defined criteria
• *Options comparative analysis* – comparison of options using evaluation metrics

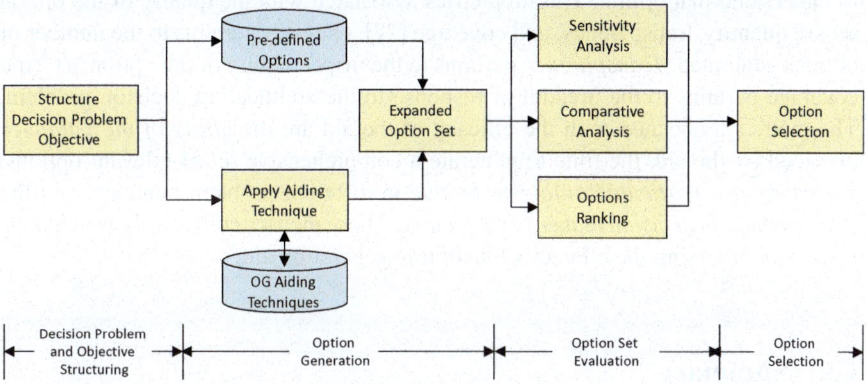

Fig. 6.3 Option generation aid usage concept

From a usage perspective, the option generation aid is an interactive application that guides and aids the user through the various stages of the option generation life cycle including objective structuring, option generation, option set evaluation, and option selection. Figure 6.3 presents the overall usage concept of the aid.

The user begins by structuring and entering the objective(s) using the templates provided by the aid. The aid retrieves pre-identified options associated with the objective and presents them to the user. The user reviews the option and decides to generate additional options using a combination of available aiding techniques. Specifically, the user chooses a technique from the menu of available aiding techniques and applies the technique to interactively generate new options. These options are added by the aid to the list of predetermined options. The user then chooses the next aiding technique and goes through the same process until satisfied with the available option set or until all techniques have been used. The user then proceeds to evaluate and rank the option set after conducting sensitivity analysis and comparative analysis of options. The user finally selects the preferred option to act on or implement.

6.5 Evaluation Metrics

Ultimately, the efficacy of the option generation approach and aid needs to be evaluated in terms of utility of the approach and usability of the aid. Evaluation metrics include the quality of the options generated, the quality of the overall option set, and the efficacy of the option generation aid. The submetrics associated with the quality of an option are implementability, novelty, and clarity. *Implementability* pertains to the realizability of the option in the real world. *Novelty* implies an unprecedented option or unprecedented use of an existing architectural option. *Clarity* pertains to how straightforward and easy it is to communicate an option to others and for others to understand that option. The submetrics associated with the quality of the option set are quantity, transparency, and coverage [27]. *Quantity* pertains to the number of options generated. *Transparency* pertains to the inspectability of the option set. And coverage pertains to the breadth of response to the architecting decision problem. The metrics associated with the efficacy of the aid are the *utility of the guidance* provided by the aid, the time to generate a comprehensive set of relevant options, the *value of a particular aiding technique* in different problem contexts, and the *improvement in decision maker performance.* These metrics collectively provide the means for assessing the efficacy of an option generation aid.

6.6 Summary

In today's highly interconnected, competitive world, traditional methods that pursue incremental improvements no longer suffice. Rather, creativity and relentless innovation are needed. These require effective methods to stimulate novel option generation using new ways of thinking. Such is the case when it comes to systems development. Option generation, a central component of decision making, is key to generating new perspectives and insights during systems architecting. Despite this recognition, it is only now that option generation is being given the attention it deserves [28, 29]. A key reason is that once routine options are readily identified, it is all too tempting to stop there. Another reason is that the generation of novel options is a substantive task that requires domain knowledge and unconventional thinking that often go beyond traditional disciplinary boundaries and problem constraints. To this end, a set of psychological and systems thinking principles are presented. These principles are intended to stimulate novel option generation especially during the early stages of systems development. A framework for an option generation aid is also presented, along with metrics for assessing the efficacy of the aid, the quality of the options generated, and the comprehensiveness of the option set. As systems continue to grow in scale and complexity, novel option generation will play a central role to achieving desired outcomes and mitigating risks in complex systems development.

References

1. Keller LR, Ho JL. Decision problem structuring: Generating options. IEEE Transact Syst Man Cybernet. 1988;18(5):715–28.
2. Raiffa H. Decision analysis: introductory lectures on choices under uncertainty. 1968. MD computing: computers in medical practice. 1993;10(5):312.
3. Janis IL, Mann L. Decision making: a psychological analysis of conflict, choice, and commitment. New York: Free Press; 1977.
4. Altshuller G. 40 Principles: TRIZ Keys to Technical Innovation. Worcester: Technical Innovation Center. Inc; 1998.
5. Fischhoff B, Slovic P, Lichtenstein S. Fault trees: sensitivity of estimated failure probabilities to problem representation. J Exp Psychol Hum Percept Perform. 1978;4(2):330.
6. Adams JL. Conceptual blockbusting: a guide to better ideas: basic books. Cambridge: Perseus Pub; 2001.
7. Arbel A, Tong RM. On the generation of alternatives in decision analysis problems. J Oper Res Soc. 1982;33(4):377–87.
8. Madni AM, Brenner M, Costea I, MacGregor D, Meshkinpour F. Option generation: Problems, principles, and computer-based aiding. Proceedings of the 1985 International Conference on Systems, Man, and Cybernetics, Tuscon; 1985.
9. Hari A, Weiss M ICDM–An inclusive method for customer driven conceptual design. Proceeding of the 2nd ASI annual total product development symposium, Pomona; 1996.
10. Dickerson C, Mavris DN. Architecture and principles of systems engineering. Oxford: CRC Press, Taylor and Francis, Auerbach Press, Abington; 2016.
11. Johnson JG, Raab M. Take the first: option-generation and resulting choices. Organ Behav Hum Decis Process. 2003;91(2):215–29.
12. Klein G, Wolf S. The role of leverage points in option generation. IEEE Trans Syst, Man, Cybern, Part C (Appl and Rev). 1998;28(1):157–60.
13. Klein G, Wolf S, Militello L, Zsambok C. Characteristics of skilled option generation in chess. Organ Behav Hum Decis Process. 1995;62(1):63–9.
14. Madni AM. Generating novel options during systems architecting: psychological principles, systems thinking, and computer-based aiding. Syst Eng. 2013;17(1):1–9.
15. Wilkins A, Baddeley A. Remembering to recall in everyday life: an approach to absentmindedness. Practical aspects of memory, vol. 1. London: Academic Press; 1978. p. 27–34.
16. Harris JE. External memory aids. Practical aspects of memory. 1978;1:172.
17. Smimov A, Kashevnik A, Levashova T, Pashkin M, Shilov N. Situation modeling in decision support systems. 2007 International Conference on Integration of Knowledge Intensive Multi-Agent Systems. IEEE; 2007.
18. Gettys CF, Fisher SD. Hypothesis plausibility and hypothesis generation. Organ Behav Hum Perform. 1979;24(1):93–110.
19. Pitz GF, Sachs NJ, Heerboth J. Procedures for eliciting choices in the analysis of individual decisions. Organ Behav Hum Perform. 1980;26(3):396–408.
20. Kahneman D, Tversky A. Subjective probability: a judgment of representativeness. Cogn Psychol. 1972;3:430–54.

21. Allport GW, Postman L. The psychology of rumor. New York: Henry Holt; 1947.
22. Holyoak KJ, Thagard P. Mental leaps: analogy in creative thought. Cambridge: MIT press; 1996.
23. Madni AM. Transdisciplinarity: reaching beyond disciplines to find connections. J Integr Des Process Sci. 2007;11(1):1–11.
24. Madni A, Moini A, Madni C. Cross-Cultural decision making training using behavioral game-theoretic framework. Advances in Cross-Cultural decision making. Informa UK Limited; 2010. p. 385–94.
25. Madni AM, Allen K. Systems thinking-enabled real options reasoning for complex socio-technical systems programs. Conference on Systems Engineering Research Redondo Beachr, CA; 2011.
26. Wang T, De Neufville R, editors. Real options "in" projects. real options conference, Paris: Citeseer; 2005.
27. Fischhoff B, MacGregor D. Calibrating databases. J Am Soc Inf Sci. 1986;37(4):222.
28. Ward P, Torof J, Whyte J, Eccles DW, Harris KR. Option generation and decision making in critical-care nursing. Proceedings of the human factors and ergonomics society annual meeting. SAGE Publications; 2010.
29. Ward P, Suss J, Eccles DW, Williams AM, Harris KR. Skill-based differences in option generation in a complex task: a verbal protocol analysis. Cogn Process. 2011;12(3):289–300.

Chapter 7
Human Performance Enhancement

"It is no use saying "we are doing our best." You have to succeed in doing what is necessary."

– Winston Churchill, former British Prime Minister

7.1 A Key Recognition

Human performance aiding and training have been viewed historically as distinct approaches to human performance enhancement. But are they really? I view these approaches as complementary and essentially lying on a continuum with aiding at one end and training at the other end. This vision was realized with the advent of the Sharable Content Object Reference Model (SCORM) standard, which has contributed to making content portable and potentially "repurposable" [1]. Thus, the SCORM standard is a key enabler of convergence between performance aiding and training. This chapter presents my methodology for integrating aiding and training using Sharable Content Objects (SCOs) to repurpose content for both performance aiding and training. Thus, the introduction of the concept of repurposable content made the convergence of human performance aiding and training possible with the approaches at opposite ends of a human performance enhancement continuum. This concept is best explained through an example scenario that along with disruptive events provides the context for developing an integrated aiding-training system [1].

Imagine a new member of the military shows up at a particular location to replace a predecessor in keeping with the personnel rotation cycle. Invariably, when the new person arrives, the predecessor is long gone. It is in this challenging situation that the combination of aiding and training can be exploited to rapidly bring the newcomer up to speed. *Aiding*, which occurs during task performance, produces a certain amount of "learning from doing" and eventually becomes integral to task performance. *Training*, which can occur anytime, can get the newcomer to the point where he/she can act with greater autonomy by relying on acquired knowledge and skills, and not on the guidance of a coach or an aid, to perform effectively on the job [1, 2]. In certain situations, however, an aid can serve as an effective needs assessment mechanism to identify knowledge and skill gaps. These gaps, in fact, can be filled through proper training [1, 2].

A.M. Madni, *Transdisciplinary Systems Engineering*,
DOI 10.1007/978-3-319-62184-5_7

The foregoing scenario also applies to small teams. Furthermore, to maximize the value delivered by the performance enhancement system, the level of aiding and training can be "tuned" to the user skill level and demands of the operational environment. This can be accomplished by incorporating adaptive aiding and training capabilities in the framework [1].

7.2 Road to the Present

The literature on aiding versus training spans approximately three decades. One of the earliest attempts to convert a consultation (i.e., aiding) system to a tutoring system was undertaken at Stanford University [3, 4]. These investigators, who began with a review of intelligent tutoring systems such as SOPHIE [5], WHY [6], WUMPUS [7], and GUIDON [4], came to the conclusion that knowledge representation was at the heart of the problem. Specifically, they identified the benefits of knowledge representations: (a) flexible multilevel, multiperspective representation; (b) capture "importance" [5], model prerequisite and generalization relationships [7], exploit "analogies," [7]; and (c) adaptable/customizable expert performance model to model the student [8]. Equipped with these insights, Clancey et al. [3] initially began working with MYCIN, an expert consultation system. Their attempts to exploit the MYCIN knowledge base for tutoring fell short but produced a couple of key insights. They discovered that the knowledge resident in a consultation system tends to be too narrow for teaching purposes. Additionally, they discovered that the consultation system did not possess adequate explanatory power. A more detailed examination of these findings revealed that (a) much of the knowledge in consultation system tends to be compiled (i.e., implicit) and, as such, a consultation system cannot always explain its recommendations and (b) the knowledge representation in consultation systems tends to be from the perspective of the task at hand and not from the perspective of the learner who typically requires multiple perspectives of knowledge to develop a well-rounded understanding of subject matter. Along the way, Clancey et al. [3] also rediscovered the procedural versus declarative knowledge tradeoff that has also been identified by several cognitive psychologists.

Two specific findings from the aforementioned studies are particularly germane to developing an integrated aiding-training framework. First, allowance needs to be made for the fact that experts carry compiled associations in their heads because they do not think about having to justify data/hypothesis links when they occasionally make long, tentative leaps to "bridge" concepts [3]. Second, these compiled associations need to be expressed declaratively to assure flexible use and redundant storage of intermediate steps and to enable satisfactory explanations of reasoning chains [3]. The integrated aiding-training continuum is informed by these findings.

7.3 Aiding-Training Trade-Offs

The analysis of aiding-training trade-offs has been revisited more than a few times over the years since the mid-1970s [10]. This trade-off is concerned with the relative emphasis placed on enhancing human performance through training versus through aiding. Simply stated, this is a trade-off between putting "smarts" in people and putting smarts in machines [2, 11]. This trade-off becomes increasingly important as advances in machine intelligence, low-cost simulation, and multidimensional visualization techniques make aiding an increasingly more viable alternative in many situations [12]. In general, highly trained individuals and teams require less aiding, while those with less training require more aiding. Beyond that, the key question is determining the relative emphases between training and aiding to cost-effectively and realistically satisfy task performance objectives based on operational context. In this regard, the general requirements for a computational trade-off analysis framework include the ability to:

- Predict the impact of training/aiding strategies on human performance, human-system performance, human-system integration, and mission effectiveness.
- Incorporate human considerations such as attributes, aptitudes, and abilities in the trade-offs performed.
- Incorporate manpower resource considerations (e.g., staffing requirements, staff qualifications) in the trade-offs performed.

It is equally important to understand the broader context within which training/aiding trade-offs need to be made. The broader context is defined by the relationships among requirements, design, training, and personnel allocation:

- *Requirements*, a crucial part of policy analysis, is concerned with conceptualizing what is needed in a manner that is both feasible and affordable, and that conforms with mandated legacy components that constrain design.
- *Design* which encompasses going from requirements to task execution is concerned with ensuring that human behavior and performance will satisfy task requirements.
- *Training* is responsible for creating qualified individuals capable of meeting job requirements/expectations.
- *Personnel (or human resources)* is concerned with producing and allocating the requisite number of trained individuals to satisfy job performance requirements.

Understanding and exploiting the interplay among these four considerations are key to making effective aiding-training trade-offs. For example, designers should be concerned with the "trainability" of people for those task requirements that result from the introduction of new technologies in the system. Similarly, training specialists need to be concerned with the ability to recruit candidates with specific competencies, abilities, and aptitudes. The answer to the latter question might well influence the answer to the trainability question [2].

Against the foregoing backdrop, decision making at the highest level pertains to trade-offs between mission effectiveness and lifecycle costs in satisfying mission requirements. Lower-level trade-offs and decisions "roll up" to determine the higher-level measures of interest. For example, manpower, personnel and training (MPT) requirements and their resulting costs are "driven" by early decisions in the system development life cycle. Ideally, downstream impacts need to be fed back upstream to modify, for example, design decisions that have undesirable MPT impacts. In practice, however, this seldom occurs because of the temporal and organizational separation of these issues. *Temporal separation* occurs because downstream decisions are often not pursued until upstream decisions are made. The problem is that upstream decision makers have few predictive tools to project downstream impacts of their upstream decisions, while there remains flexibility for those decisions to be reversed or modified. The *organizational separation* of issues results in "suboptimization" in the sense that each issue and associated trade-offs are resolved in a locally optimal manner, thereby undermining global optimality [13]. While some of this organizational separation is a result of historical precedents and political expedience, it is also the case that much of the organizational decomposition (and hence separation) reflects an attempt to cope with the complexity of designing large-scale systems. In this regard, what is needed are methods and tools that enable consideration of more global trade-offs while making complex designs more manageable.

Finally, there are several ways to train and aid people and, in fact, there is data available to show the benefits and limitations of each [14]. What is lacking, however, is a way of performing trade-offs among alternatives without having to perform an empirical study. This deficiency, in part, is due to the lack of an integrative framework for trade-off analyses [1, 2]. *The key to developing such a framework is incorporating the means to predict rather than merely describe and measure the impact of specific training and aiding alternatives* [1, 2].

7.4 Parameterizing Aiding-Training Continuum

It is useful to think of aiding and training in terms of multiple levels along the aiding-training continuum [1, 15]. For example, at the training end of this continuum, one can employ e-learning courseware prior to doing actual work (e.g., maintenance troubleshooting). Moving along the continuum, the next level is tutoring or just-in-time training while the job performer is engaged in doing work (e.g., delivering a quick course snippet on a "widget" that the maintainer has to repair or remove). Further along the continuum is question answering or explaining (e.g., "How many connecting bolts are there for a widget?"). The next level is aiding, which typically involves operations (or maintenance) procedures, that the integrated aiding-training system user needs to access. The use of such procedures might require explanations, tutoring, or both. Thus, users may need training when using the aiding capability. The rightmost extreme of the continuum is defined by automation which does

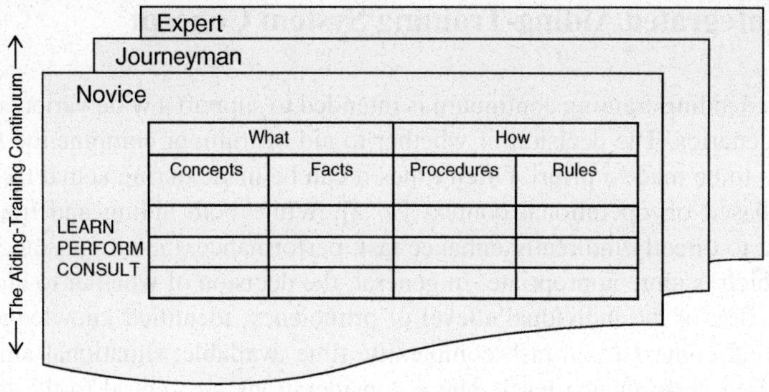

Fig. 7.1 The aiding-training continuum

not apply directly to the aiding-training continuum in that the integrated aiding-training system, as envisioned, does not cover sensing and actuation abilities.

In addition to the *types of support*, there are three *perspectives of support* that users expect from the system: What, How, and Why. "What" provides factual information via illustrations, text, and possibly audio. "How" provides the steps to perform a task via illustrations, text, and possibly audio. "Why" provides a deeper explanation of, for example, the functioning of an assembly (e.g., "Why is shear pin A needed?").

Finally, military organizations classify learners/job performers into three levels: Novice, Journeyman, and Expert. Novices need specific support with considerable handholding. Journeymen "have been there before" and usually need only a "refresher" – the system knows this information from the history of the individual stored in a learning management system (LMS). Experts tend to be highly skilled and are invariably interested in clarifying a few points, which the system needs to provide succinctly. The system knows that an individual user is an expert, journeyman, or novice based on the user's history of courses taken, performance scores, and length of experience. Such historical data is typically stored in the LMS. In today's world, people need to learn very different subjects than what they studied in school, or college, or that they used in their past jobs. What this means is that an individual could be an expert in one area and a novice in another. Thus, the role in which a person is expected to perform is the major determinant of whether that person is a novice, journeyman, or expert. The results of such an analysis, can be organized using the three-dimensional framework for structuring the knowledge base (Fig. 7.1). A fourth dimension (not shown in the figure) is Tasks. The nature of support provided by the system to users depends on their position in this framework [1, 15]. For one specific task, they may be novices in need of training, while for another task, they may be experts needing only terse responses/clarifications [16]. Such distinctions can be made by drawing upon: (a) information captured in the user's "electronic jacket" or LMS; and (b) the nature of the user's query [12].

7.5 Integrated Aiding-Training System Concept

Integrated aiding-training continuum is intended to support a wide variety of operational scenarios. The decision of whether to aid or train, or combine the two does not have to be made a priori. Often times it can be made during actual task performance based on operational context [1, 2]. While both aiding and training are intended to directly/indirectly enhance task performance, the circumstances determine which is more appropriate. In general, the decision of whether to aid or train is a function of the individual's level of proficiency, identified knowledge gap(s), operational context (e.g., task complexity, time available, situational ambiguity), and desired performance level. These considerations are central to the integrated aiding-training system concept.

A key element of the integrated aiding-training concept is the "electronic continuity folder," which provides the incoming job performer access to the predecessor's activities, work products, assumptions, and other relevant "leave behind" materials [1]. In this regard, the system queries the newcomer to determine whether or not the newcomer has read and assimilated (not necessarily accepted) the lessons from the predecessor. The newcomer, in turn, can update the knowledge (content) in the electronic continuity folder with his/her experiences (including lessons learned). Updates to the body of knowledge contained in the electronic folder are transmitted (i.e., sent back) to the schoolhouse and other distributed training establishments to ensure up-to-date training content at these sites. The updated training content can then be used to develop new/enhanced training for recruits who ultimately show up in the operational theater at some point in the future. In this regard, a key goal is to ensure that the contents in the electronic continuity folder and at the schoolhouse and other training establishments are identical and current. As noted earlier, content repurposing is central to the system. Content (re)purposing consists of augmenting individual sharable content objects (SCOs) from the content repository with learning (or performance) objectives, instructions, and assessment (or performance support) strategies and presentation "objects." Figure 7.2 presents the content repurposing approach.

As shown in Fig. 7.2, the SCOs, once authored by subject matter experts (SMEs), are persistently stored in the content repository, whose contents can be conveniently accessed and (re)purposed [17]. The purposing mechanism for e-learning takes one or more SCOs, augments them with learning objectives (LOs) and instruction and assessment strategies (IAs), and presents them in the form of media objects (MOs). Similarly, the purposing mechanism for performance support/job aiding takes one or more SCOs, augments them with performance (i.e., aiding) objectives (POs) and aiding and assessment strategies (AAS), and then presents them in the form of media objects (MOs). Appropriately purposed content is provided by the integrated aiding/training system to the user during learning or task performance.

Key Tenets The fundamental tenets of integrated aiding and training are presented in Table 7.1.

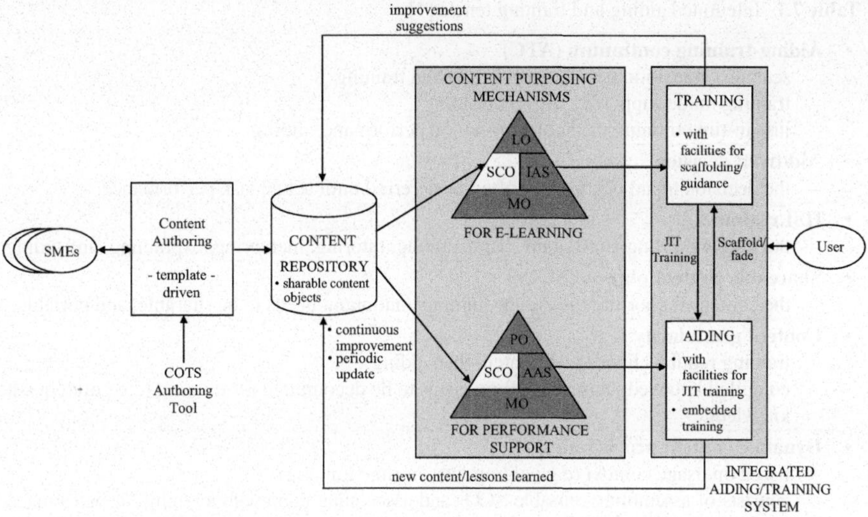

Fig. 7.2 Content (Re)purposing

7.5.1 User-System Interaction

In creating an integrated training and aiding system, it is necessary to achieve consistency in user-system interaction and user interface design. Such consistency is key to keeping the user "cognitively coupled" with the system across the aiding-training continuum. Our parameterization of this continuum is through a carefully selected set of functions (i.e., LEARN, PERFORM, CONSULT) that allow the user/operator to seamlessly move between aiding and training:

(a) LEARN offers the user formal web-based courseware. It typically applies in environments that are not time-stressed. The user at any point can move to PERFORM or CONSULT as the need arises. The user will typically move from LEARN to PERFORM to ascertain his/her ability to perform what was learned. Similarly, the user will move from LEARN to CONSULT to acquire information through peer-to-peer collaboration, subject matter expert (SME) consultation, community of interest (CoI) participation, and review of reference material stored in databases or accessible over the web through URLs.
(b) PERFORM allows the user to apply the material learned in a simulated or real operational context. The user can receive on-demand performance support during PERFORM. At any point during PERFORM, the user might discover that he/she lacks specific knowledge required for task performance. In this case, the user can move to LEARN and receive just-in-time training or to CONSULT and receive advice, recommendation, or guidance from remote SMEs and community of interest (CoI) or acquire requisite information from databases or web sources.

Table 7.1 Integrated aiding and training tenets [1]

• **Aiding-training continuum (ATC)** – seamless transition between job aiding and training – training mode supported with scaffolding – just-in-time training available during job performance aiding
• **Aid/train decision** – the decision to aid or train can often be deferred until actual task performance
• **HSI regime** – the training-aiding continuum spans training, tutoring, mentoring, explaining, and aiding
• **Shareable content objects (SCOs)** – the content supporting the aiding-training continuum needs to be sharable, and portable
• **Content granularity** – training requires finer grain content than aiding – compiled knowledge used in aiding needs to be decompiled for training (e.g., prerequisite knowledge)
• **Dynamic content provisioning** – based on usage context (e.g., aiding, just-in-time training) – consists of assembling reusable SCOs and associating them with learning (or performance support) objectives and corresponding instruction (or performance support) strategies
• **User interface consistency** – common user interface in terms of information location on the screen and user-system interaction
• **Guided navigation** – training and aiding navigation sequences differ – training navigation: concept, function(s), manipulated entities, assets – aiding navigation: "how to" knowledge, manipulated entities, function(s), concept/how to
• **Metadata tagging** – full support for SCORM/IEEE learning object metadata (LOM) – customizable, extensible metadata – metadata templates to facilitate tagging
• **Content authoring/packaging** – SCORM-compliant content package generation (IMS XML manifest, etc.) – graphical language (navigation and sequencing, describing course content)

(c) CONSULT provides the user access to a variety of information sources (e.g., human experts, CoI members, reference documents, and web sources). It can be invoked from both LEARN and PERFORM. The characterization of user activities as LEARN, PERFORM, and CONSULT reinforces the types of activities that the system supports along the human-system integration continuum. It is also important to realize that learning continues to occur in the PERFORM and CONSULT activities as well. Once the user has employed LEARN, PERFORM, and CONSULT a few times, the user will become quite accustomed to moving smoothly between these three modes.

7.5.2 *Typical Usage Scenario*

The user can engage in any one of the three activities: LEARN, PERFORM, and CONSULT. These activities span the human-system integration continuum. A typical system usage concept is described next.

After log-in, users can indicate their initial purpose. The choices available are to specify a learning objective and take a formal course, perform a task with appropriate job aiding and performance support, or consult a variety of information sources (e.g., community of interest, virtual classmates, subject matter experts, or reference material in databases or accessible over the web). Once logged into the system, users can move seamlessly between learning, performing, and consultation. If users select LEARN, they get access to a variety of e-learning courses in the form of topics and lessons. During LEARN, they always have the option to select PERFORM or CONSULT.

Upon selecting a learning objective, the user is presented with the corresponding lesson. The lesson is supported with semantic icons that cue the user about lesson content. For example, the wrench icon denotes assembly/disassembly, and the puzzle piece implies problem solving, while the light bulb implies something to remember or recall. The lesson is augmented with "quick links" to reference material allowing the user to rapidly access job-/task-related material while in the learning mode. Lessons have embedded thumbnail images that can be expanded simply by clicking on them.

Whether performing practice exercises or taking tests, the user performs drag-and-drop operations, and responds to fill-in-the-blank prompts, and/or multiple choice questions. During exercise taking, the user gets feedback such as "correct" or "wrong," followed by appropriate remediation. However, during test taking, users are on their own, i.e., receive no feedback. Thus, typical exercises and tests take the form of drag-and-drop operations and "fill-in-the-blank" prompts or answer multiple choice questions.

In the PERFORM mode, the user receives performance support. Performance support can take a variety of forms. The user can take advantage of thumbnails which show the type of content available and then search top-down (system-subsystem-component) or bottom-up (component-subsystem-system) based on the user's purpose. The user can expand a search result to view details such as schematics, text descriptions of procedures, videos, and 3D animations. The user always has a link back to the lesson material.

The CONSULT mode provides the user access to reference material, parts library, catalogs, CoI, and live chats or email communication with experts. Finally, the system maintains a history of a user's prior sessions and uses that knowledge to deduce specific concepts or procedures that are challenging for a particular user. Based on such findings, the system recommends specific refresher lessons to the user.

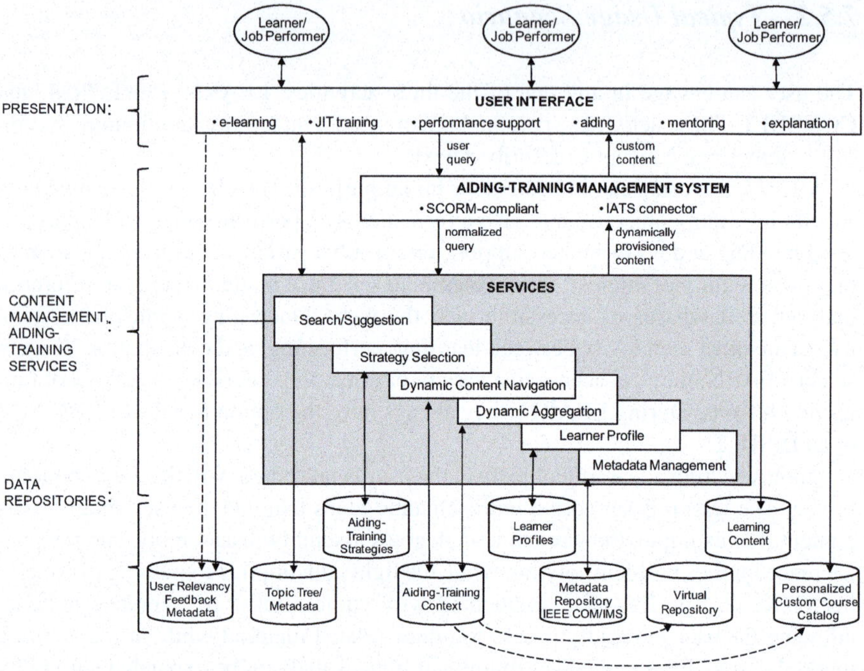

Fig. 7.3 Integrated aiding-training system functional architecture

7.5.3 *Architecting and Implementation*

The system is architected as a standards-compliant, scalable, and extensible set of services (Fig. 7.3). Central to the system is the dynamic provisioning of content based on user needs (e.g., aiding, mentoring, training).

In the following paragraphs, the key characteristics of the architecture are described:

- *Standards Compliance*. Compliance with standards is critical to ensuring backward compatibility and interoperability with the existing learning management system (LMS) infrastructure (i.e., portal products and learning content management tools) as well as the existing courseware.
- *Shareable Content Object Reference Model (SCORM) Framework*. The SCORM 2004 4th Edition requires modularity in content development. Modularity allows interrelated learning resources to be grouped into aggregate entities called "learning objects." Learning objects can be developed relatively independently from one another and made accessible via an online web content repository. The independently developed, distributed learning objects can be reused and repurposed across courses, disciplines, training programs, and organizations. For the system, there is an additional requirement – the ability to dynamically aggregate learning objects into a coherent and effective learning experience.

- *Automation*. Courseware development approaches that are SCORM-compliant largely employ a manual development process that involves a priori aggregation of learning content. The aiding-training system overcomes this limitation by providing the capability to automatically and dynamically aggregate learning objects into context-specific, pedagogically grounded, personalized learning units. This type of "dynamic content provisioning" requires merging and reconciling relevant search results into a single learning path, sequencing the selected learning objects along the path, and linking the selected learning objects into an organized, coherent instructional unit. Of necessity, dynamic provisioning in the system needs to be based only on those parameters that are available when the learning session begins. Examples of such parameters are the keywords in the learner's query and the desired level of aiding/training. The query is typically based on learning objectives, or the learner's specific interest(s).
- *Service-Oriented Architecture*. Service-oriented architecture (SOA) offers an appropriate construct for implementing the system in that it exploits the full range of capabilities offered by the web services stack (e.g., service description, registration, discovery, composition, security, policy). The system architecture comprises a set of cooperating Web Service Description Language (WSDL) services including a Search/Suggestion Service, a Dynamic Content Aggregation Service, a Dynamic Content Navigation Service, a Learner Profile Service, and a Metadata Management Service. The functionality of each service is described below:
  - *The Search/Suggestion Service* leverages extended SCORM/IEEE LOM metadata, indexes SCORM metadata and content, and generates relevancy-ranked sets of learning content in response to user queries.
  - *The Strategy Selection Service* employs contextual information to determine the most appropriate form of intervention along the aiding-training continuum.
  - *The Dynamic Aggregation Service* consists of two constituent mechanisms: (1) a path generation mechanism, which brings together search results that are most closely related to the user query, into a coherent path, as defined by relationships between metadata values appearing in the learning objectives metadata, and (2) a sequencing mechanism, which orders the objects along the path, in a manner consistent with the learning topic sequence and the instructional role sequence for the learning unit/lesson being worked. The Aggregation Service also adds objects retrieved by the sequencing mechanism to the current learning object organization.
  - *The Dynamic Content Navigation Service* inserts the learning object organization into the IMS Manifest structure; displays the dynamically provisioned content to the user; processes navigation requests, including requests to bookmarked lessons and suspension of the course; and communicates with the Learner Profile Service to store tracking data.
  - *The Learner Profile Service* enables users to modify personal preferences and attributes. It updates session tracking data such as the number of times a

learning object has been viewed outside a course, configured into a course, bookmarked, or played as part of a course.
  - *The Metadata Management* Service provides access to SCORM metadata associated with each learning object.

- *Dynamic Content Provisioning*. Dynamic content provisioning is concerned with purposing content "on-the-fly" in response to the user's learning (or job performance) needs. It is achieved through a number of components that cooperate to deliver contextualized, pedagogically grounded, personalized content. The dynamic content provisioning components currently implemented in the system include a metadata search engine, a learning objective selector, a dynamic object aggregator, and a query processor.
- *Content Model*. The system's content model contains the learning content. The system architecture separates the learning content from the dynamic aggregation, sequencing, and navigation logic. This separation ensures reuse of learning objects (LOs) because they are not bound to a particular course package implementation or organization. This aspect of the system architecture also allows content developers to tag and annotate LOs by providing descriptive metadata characterizing both technical content and pedagogical aspects of the LOs. This information is useful to courseware developers in selecting LOs. This information can also be used by the system suggestion service to recommend the most appropriate learning content when there are several candidate LOs available to satisfy a learning objective.

Finally, backward compatibility is provided, where needed, by the system architecture where needed. Since the new crop of learners and job performers tends to be digitally-savvy, they can adapt to the user interface. This implies that backward compatibility in user interface design and user-system interaction is not essential. However, backward compatibility with respect to the system infrastructure and reuse of LOs is highly desirable. To maintain backward compatibility with the existing LMS infrastructure and to ensure reuse of the LOs outside the system environment, the aiding-training system architecture leverages standards-based metadata to characterize its content model. The use of standards also allows existing courseware content to be imported and repurposed within the system environment assuming that such content also has a compatible metadata model.

7.5.4 Content Authoring

Content authoring requires a tool that enables content creation, tagging, packaging, and metadata lifecycle management associated with adaptive, reusable content objects.

Within the aiding-training system environment, the metadata allows learning resources to be searched, discovered, and recommended within the context of the aiding-training continuum. Learning resources, intended to be reusable, are generally

Table 7.2 Authoring of metadata categories

• **General** – information about the resource or content aggregation
• **Lifecycle** – history and current state of the resource and those who have contributed to its development (optional for assets)
• **Meta-metadata** – specific information about the metadata record itself, as opposed to the resource described by the metadata
• **Educational** – key educational or pedagogical characteristics of the resource (optional for all resource types)
• **Relation** – relationship between the resource and other resources (optional for all resource types)
• **Annotation** – comments on the educational use of the resource, generally by parties other than the original author (optional for all resource types)
• **Classification** – where the resource is placed in relation to a specified taxonomy or classification scheme (optional for assets) The classification category is intended to describe how the resource fits into one or more standardized classification systems or taxonomies

tagged with metadata to facilitate search and discovery across content repositories. The system metadata authoring tool supports two broad classes of metadata: (1) context-independent (immutable) metadata, which are basically the equivalent of library catalog-type entries, and (2) context-dependent metadata, which are only meaningful with respect to a particular "context" and can be defined or redefined dynamically based on the learning objectives and aiding-training strategies. Learning object metadata annotations provide information about learning objects including reuse potential for aiding and/or training. Table 7.2 shows the authoring of several broad categories of metadata at different levels in the content hierarchy.

7.5.5 System Evaluation Criteria

System evaluation can be conducted in stages. The key evaluation factors are usability/acceptability, validity, and viability [18] (Table 7.3).

The validity, acceptability, and viability can be addressed within the context of the application's domain. For example, for the maintenance application domain, the content went through extensive evaluation and refinement at the Intermediate Maintenance Activities in San Diego and Mayport. Both end users and qualifiers worked intensively with the system over a 3-day period. Users found the system easy to work with. Qualifiers concluded that the system would fit naturally within their existing training activities while relieving them to attend to other chores. The viability of the system was established in terms of cost savings resulting from avoidance of travel and boarding needs and time away from work that is required today. An independent study conducted by the responsible authority at the Center, Fleet Forces Command (CFFC) showed substantial cost savings from $1172 per seat per year to $28 per seat per year for this course. The substantial cost savings resulted

Table 7.3 Evaluation factors

• **Validity**: Does the solution solve the problem in a technically valid manner. For example, is the technical content correct and sufficient for solving the maintenance problems addressed
• **Acceptability**: Does the solution fit into the way things are or could be done both individually and organizationally. For example, is system usable by the targeted personnel and organizationally supportable
• **Viability**: Are the benefits of the solution greater than the costs in terms of acquisition and recurring costs as well as costs of change

from avoidance of travel and boarding costs, instructor costs, and classroom facility costs. This calculation was based on $620 K turnkey cost of the software, and a 3700 crew member complement that was required to operate the ship at sea. These results were also based on surface ship watertight closure training requirements specified in CIN K-495-0401.

7.6 Summary

Human performance aiding and training are complementary approaches to human performance enhancement. Despite this recognition, the two approaches have not been brought together within a common rubric because they were pursued by two different communities and because there are subtle differences between the two. For example, information that is implicit in a performance aiding system often needs to be made explicit in a training system [1]. Also, in general, the time available for training and aiding tends to be different [1]. However, with the emergence of just-in-time training and on-demand performance support, these differences have begun to blur [1]. With the advent of the SCORM framework in early 2000, it became a requirement for content to be in the form of shareable content objects (i.e., SCOs); it has become possible to aggregate and disaggregate content, as well as (re)purpose content from aiding to training, and vice versa. The *aiding-training framework* presented in this chapter is an example of disciplinary convergence (i.e., the aiding-training continuum) between aiding and training approaches that have different disciplinary roots. Aiding has its roots in decision science and artificial intelligence, while training has its roots in psychology. Building on this disciplinary convergence, it has become possible to create new capabilities. For example, the integrated aiding-training framework facilitates content repurposing in support of the aiding-training continuum. While the SCORM framework specifies only a priori aggregation/assembly of content to satisfy learning objectives, the aiding-training framework extends this capability through dynamic content provisioning. The latter provides the necessary mechanisms for on-the-fly content (re)purposing. The prototype system presented in this chapter provides a unified user interface based on the "LEARN-PERFORM-CONSULT" construct, which enables seamless traversal of the aiding-training continuum. The system employs a service-oriented architecture

(SOA) with the core services being Search/Suggestion Service, Dynamic Content Aggregation Service, Dynamic Content Navigation Service, Learner Profile Service, and Metadata Management Service.

In summary, by organizing the various forms of human performance and productivity enhancement methods along a human performance enhancement continuum, and providing a common user interface, users no longer have to learn how to work with multiple software systems. They no longer have to experience the jarring effects of discrete mode switching from performance aiding to training, and vice versa. These characteristics are key to end user acceptance [16] and eventual proliferation of this capability. The creation of this capability was the result of recognizing and exploiting the growing convergence between decision science, artificial intelligence, and psychology. This capability today can be invaluable for both the defense and commercial sectors, where affordability, adaptability, and ease-of-learning are highly valued.

References

1. Madni AM. Towards a generalizable aiding-training continuum for human performance enhancement. Syst Eng. 2011;14(2):129–40.
2. Rouse WB, Johnson WB. Computational approaches for analyzing tradeoffs between training and aiding. DTIC document. 1990.
3. Clancey WJ, Letsinger R. NEOMYCIN: Reconfiguring a rule-based expert system for application to teaching: Department of Computer Science, Stanford University; 1982.
4. Clancey WJ. From GUIDON to NEOMYCIN and HERACLES in twenty short lessons. AI Mag. 1986;7(3):40.
5. Brown JS, Burton RR. Multiple representations of knowledge for tutorial reasoning. Representation and Understanding: Bobrow, DG and Collins, AM, Eds., Academic Press, 1975. p. 311–349.
6. Stevens AL, Collins A. Multiple Conceptual models of a complex system. DTIC document. 1978.
7. Goldstein I. Developing a computational representation for problem solving skills. Cambridge: Massachusetts Institute of Technology, Artificial Intelligence Laboratory; 1978.
8. Brown J, Burton R. Diagnostic models for procedural bugs in basic mathematical skills. Cogn Sci. 1978;2(2):155–92.
9. Riedl MO, Young RM. An objective character believability evaluation procedure for multi-agent story generation systems. Intelligent virtual agents. New York: Springer Science + Business Media; 2005. p. 278–91.
10. Madni AM. Game-based simulation for cross-cultural decision making training. Hum Factors Man. 2012;23(2):85–94.
11. Sage AP. System design for human interaction. New York: IEEE Press; 1987.
12. Madni AM. AgileTecting™: a principled approach to introducing agility in systems engineering and product development enterprises. J Integr Des Process Sci. 2008;12(4):49–55.
13. Rouse WB. Training and aiding personnel in complex systems. Manprint: Springer Science + Business Media; 1990. p. 417–32.
14. Booher HR. Handbook of human systems integration. Hoboken: John Wiley & Sons; 2003.
15. Rouse WB. People and organizations: explorations of human-centered design. New York: Wiley; 2007.

16. Madni A. The role of human factors in expert systems design and acceptance. Hum Factors: J Hum Factors Ergon Soc. 1988;30(4):395–414.
17. Madni AM, Madni CC. Context-driven collaboration during Mobile C2 operations. Proceedings of the society for modeling and simulation international. 2004:18–22.
18. Rouse WB. Design for success: a human-centered approach to designing successful products and systems: New York, Wiley-Interscience; 1991.

Chapter 8
Design Elegance and Systems Engineering

"A designer knows he has achieved perfection not when there is nothing left to add, but when there is nothing left to take away."

– Antoine de Saint-Exupery, French aviator

8.1 Elegance in Design

The term "elegance" is typically associated with aesthetics. In complex systems design, elegance is what separates the merely functional from the truly engaging. An elegant design is often inspired by a theme that drives its creation. It engages users and encourages exploration. The process of elegant system design is a creative process that focuses on the total experience and exploits systems thinking, probing and questioning, and the use of appropriate analogies and metaphors to simplify system architecture and design by eliminating extraneous constraints that can show up because of poor problem framing.

Design require synthetic (i.e., integrative) problem solving coupled with creativity especially in problem framing, and relatively open solution paths. Creative designs share a few common characteristics: *novelty* (i.e., new or unprecedented characteristics), *relevance* (ability to satisfy problem constraints), and *effectiveness* (ability to satisfy performance and nonfunctional requirements). Such designs exhibit *functional creativity*. However, a creative solution to a design problem goes beyond functional creativity and into the realm of *design elegance*. An elegant design, while associated with high levels of functional creativity [1], also circumvents extraneous problem constraints that typically increase the number of system trade-offs. An elegant design invariably results from a new way of looking at a known problem that reduces the number of constraints that need to be satisfied. Occasionally, it is associated with a creative solution to a previously unrecognized need.

Elegance has also been characterized as an effortless display of sophistication. Gelernter [2] suggests that elegance is an effective marriage of simplicity and power. O'Connor [3] claims that an elegant design is correct, efficient, and "pleasing to contemplate." Alberti's characterization of beauty is also an apt characterization of elegance. He characterizes beauty as "the adjustment of all parts proportionately so that one cannot add or subtract or change without impairing the harmony of the whole." The term "elegance" is increasingly being associated with system designs

A.M. Madni, *Transdisciplinary Systems Engineering*,
DOI 10.1007/978-3-319-62184-5_8

that are simple and effective. According to Billows [4], for a complex system design to be considered elegant, the system must function according to its stated purpose, and the "design pressures constraining the system design must be simultaneously relieved." Lewis [5] has a slightly different take on elegance when he states that "... simple answers provide elegant insights to complex problems." The concept of elegance has also been addressed in object-oriented software design [6].

The pursuit of *elegance in design* can be traced back to earlier times. In 25 BC, Vitruvius defined elegance as creating pleasing structures to the eye, resilient to the forces of nature, meeting the needs of the user, and resulting in a harmonious combination of design, art, and material [7]. Owen Edwards observed that elegance is unique in the way it evokes an emotional reaction [8]. Rubinstein and Firstenberg [9] define an elegant design as one that takes a creative approach to coping with complexity. Gell-Mann [10] expands on the notion of complexity by noting that "the length of the most concise description of an object can determine its effective complexity." Gell-Mann adds that "these concise descriptions of objects as schemas are created by condensing observations of the regularities in the world around us" [4]. Ockham's razor is a well-known principle that recommends selecting the hypothesis that makes the fewest assumptions from among competing hypotheses [11]. Others identify creativity as a means to ameliorate complexity [12–14]. Meyer and Lehnerd [15] define an elegant system as "one in which all subsystems, taken as a whole, create the greatest output performance for the least inputs." Brooks [16] states that "the ratio of function to conceptual complexity is the ultimate test of system design." Bellman [17] characterized elegance in design as stemming from an iterative process that begins with a wonderfully complex and unimaginably difficult solution that on the next pass is still complicated but understandable, and on the following the pass explainable but not simple, and on the final pass is so simple that it is intuitive. Kalaba [18] made an insightful observation as well when he remarked that "it is a mistake to think that if something was written (or made) simply, it was written (or made) by a novice or a simple mind." A more holistic perspective, offered by Rechtin [12], suggests that elegance is in the eyes of the beholder. More recently, Michael Griffin, former NASA administrator, characterized elegance in engineering design as "an ineluctable concept in that it is immediately apparent when it exists, and yet it is difficult to define, impossible to quantify, and so far, apparently incapable of being taught" [19]. Griffin [20] went on to define elegant design as one possessing the attributes of effectiveness, robustness, and efficiency, with minimum unintended consequences.

It is important to realize that an elegant design need not be entirely original nor does it have to be foolproof. It was none other than Pablo Picasso who famously observed, "Good artists copy, great artists steal." Apple's cofounder and visionary, Steve Jobs, proudly proclaimed that Apple has "always been shameless about stealing good ideas." He proudly announced that "part of what made the Macintosh great was that the people working on it were musicians, poets, artists, zoologists, and historians who also happened to be the best computer scientists in the world." What Jobs described are transdisciplinarians who sought to make connections among different disciplines to achieve elegance in design [21]. In a *Businessweek* interview in

1998, Jobs proclaimed that "it's really hard to design products by focus groups. A lot of times people don't know what they want until you show it to them." Jobs never deviated from his relentless focus on simplicity and power. He famously remarked that "simple can be harder than complex. You have to work hard to get your thinking clean to make it simple," he said. "But it's worth it in the end because once you get there, you can move mountains." This is the promise and potential of elegant systems design.

Regardless of one's viewpoint or definition, the means by which constraints are dissolved, reduced, or circumvented through effective problem reframing (or reformulation) are the more important considerations that go into the creation of an elegant system design.

8.2 Elegant Design Characteristics

An elegant design is an important predictor of success in the marketplace because an elegant design benefits all stakeholders (Table 8.1). While design is a mapping from function to form [22], elegance is introduced not merely in the function and form but also during this mapping and beyond to assure a total experience that is memorable and engaging. A truly elegant design is one that does away with having to make performance, or capability limiting trade-offs among desirable system characteristics. A combination of aesthetic appeal and functionality, an elegant design builds a visceral connection with the user by progressively "following" its story through the user's experience with the system. An elegant system design often embodies a capability that humans were unaware they needed and now cannot do without.

Both software and manufacturing industries offer examples of elegance. Elegant software requires a fraction of the code that exists in a poorly written program that performs the same function. Elegant software is also easier and cheaper to debug and cheaper to document and maintain because it has less code [23].

Table 8.1 An elegant design benefits all stakeholders

• **Customers/users:** like the experience of using/working with the product or system
• **Owner/seller:** reaps higher margins because product is efficient, i.e., produces more output for less input
• **Maintainer/tester:** system is easily instrumented, and easy and inexpensive to troubleshoot/debug
• **Manufacturer:** product/system is less susceptible (i.e., sensitive) to variability in the manufacturing process
• **Distributor:** system/product is easy to explain to end users and resellers
• **Inventor:** system/product exhibits the right technical fits
• **Investor:** system/product has the potential of creating or redefining an entire market

Elegant manufacturing implies a product that is less sensitive to variability in the manufacturing process and requires fewer parts and fewer part types than the norm. An elegant manufacturing process is not only a streamlined, simplified manufacturing process but also one that can potentially create or redefine an entire industry [25].

Polya defines elegance of a mathematical theorem as follows: "The elegance of a mathematical theorem is directly proportional to the number of independent ideas one can see in the theorem and inversely proportional to the effort it takes to see them." [24] According to Ackoff [26], there are three ways of responding to a design problem: *resolve*, which implies reaching an outcome that is merely satisfactory ("stop gap"); *solve*, which implies optimizing the solution to a problem in the presence of constraints ("constrained optimization"); and *dissolve*, which implies creatively removing barriers that lead to an effective solution. The latter implies circumventing constraints in a manner that precludes having to make trade-offs between desired characteristics, and is a key aspect of elegant design. Table 8.2 presents the key characteristics of an elegant system design.

An elegant design fosters an emotional bond with the human (user or operator). This phenomenon is apparent with Apple products (e.g., Macintosh, iPod, iPhone, iPad). Joy of use is more than usability, utility, and user experience. It is a visceral connection and an emotional bond between the user and the system. This bond is a result of the user's experience, system's responsiveness, system's adaptivity to the user, and system's appearance, feel, and personality that evoke certain emotions and impressions in the user. Systems that are simple, self-explanatory, and with a personality that is reflected in their design and interaction with the user are the ones

Table 8.2 Characteristics of an elegant system

Characteristic	Description
Purposivity	Accomplishes purpose or goal with minimum side effects/negative consequences, given priorities
Parsimony	Accomplish intended purpose with minimum number of components, resources, and interventions in routine/contingency situations; minimum structural complexity
Transparency	Inspectable system behavior during operation or use
Scalability	Design scales linearly with increase in complexity
Resilience	Perform effectively in the face of anticipated and unanticipated change
Bonding	Visceral/emotional connection with users/operators
Efficiency	Accomplish desired outcome with minimum resources, effort, waste
Evolveability	Adapt/extend seamlessly to meet new market/operational demands
Affordability	Total costs within customer's acceptance threshold
Usability	Ease of use with negligible errors and error rate
Utility/Impact	Monetary/nonmonetary net positive effect achieved for investment made
Predictability	System behavior can be determined in a variety of future contexts and scenarios
Learning	Learn continuously from interactions with users and the environment
Simplicity	Minimize number of applicable trade-offs through proper problem framing
Experience	Pleasurable feeling from the point of acquisition/purchase through usage, and post-usage

that create joy of use. Joy of use translates into customer/user loyalty and success in the marketplace. Ultimately, it is about the total experiences with the system that is much more than just design elegance. And this outcome requires systems thinking.

From a complexity perspective, there are two types of elegance: systemic elegance and perceived elegance. *Systemic elegance* thrives on simplicity through minimalist thinking and parsimony. *Perceived elegance* hides systemic or organizational complexity from the user, thereby reducing the user's cognitive and perceptual load.

8.3 Elegant Solutions

Lateral (or associative) thinking can potentially lead to flashes of insight, especially when one actively looks to other disciplines to identify appropriate parallels and analogies. Lateral thinking can often produce approaches that dissolve performance or capability limiting constraints associated with conventional formulation of a design problem. An architect or designer with a multidisciplinary bent is more likely to explore nontraditional approaches and thereby acquire novel insights that can often serve as the basis for elegant solutions [21, 27]. An elegant solution is one that simultaneously relieves "design pressures" on a system so that all objectives are met without having to make compromises [4]. Design pressures arise when the system is required or expected to function at levels not achievable in the operational environment by the existing design. In this sense, design pressures can be viewed as failure of a system to meet operational requirement/constraints. However, viewing elegance as merely the balancing of constraints is too narrow a view. Elegance is circumventing the need to engage in having to make compromises (in response to design pressures) while achieving objectives. In fact, there are several examples of elegant solutions that can be traced back to design pressures [4]. The geodesic dome, for instance, distributes the structural forces so well that the structure is made from minimum materials and is of minimum weight [28].

Another example of an elegant solution that involved a new way of looking at a known problem can be seen in the work of a group of civil engineers who were confronted with the necessity of drilling a large number of holes in extremely hard concrete slabs so that the slabs could be bolted together. The drill bits kept on snapping, and the problem was initially viewed as one that required designing tougher and sharper bits. It was only when the objectives were revisited and the problem reformulated that the task was redefined as one of avoiding holes altogether. It was only then that an elegant solution was found. The elegant solution in this case resulted from a reframing of the problem from one of having to drill holes to one of creating interlocking concrete slabs.

A commonly encountered system problem is the need to perform trade-offs among, highly desirable system attributes that are viewed as competing attributes. Often times, the need for the trade-offs is a direct result of suboptimal problem framing or formulation. The following example illustrates this concept. A market research firm needs the average age and average salary of a special interest group. The firm needs to protect the privacy of the group (i.e., cannot ask for actual age and actual salary). An elegant solution to this problem is one achieves the objective of the market research firm without violating the privacy constraint. How can such an elegant solution be achieved?

An elegant solution to this problem is as follows.

Each group member *i* adds a random number r_i to his/her salary, S_i, to produce Y_i:

$$Y_i = S_i + r_i \qquad i = 1, \ldots n$$

The first group member adds a random number r_1 to salary S_1, and shares the result with the next member of the group.

$$\text{Thus, } Y_1 = S_1 + r_1$$

The next group member adds a random number to his/her salary and adds the result to the random number Y_1, and so on. Thus,

$$Y_2 = Y_1 + S_2 + r_2 = S_1 + r_1 + S_2 + r_2 = \sum_{2}^{i=1} S_i + \sum_{2}^{i=1} r_i$$

$$\vdots$$

$$Y_n = Y_{n-1} + S_n + r_n = \sum_{n}^{i=1} S_i + \sum_{n}^{i=1} r_i$$

$$Y_n = \sum_{n}^{i=1} S_i + \sum_{n}^{i=1} r_i = \text{sum of all salaries} + \text{sum of all random numbers}$$

Now the first group member takes Y_n and subtracts the random number that he/she had added previously.

$$\text{Let } Z_1 = Y_n - r_1$$

Then,

$$Z_2 = Z_1 - r_2 = (Y_n - r_1) - r_2 = Y_n - r_1 - r_2$$

$$Z_3 = Z_2 - r_3 = Y_n - r_1 - r_2 - r_3$$

$$\vdots$$

$$Z_n = Z_{n-1} - r_n = Y_n - r_1 - r_2 - r_3 \ldots - r_n$$

Finally,

$$\text{average salary} = \frac{Z_n}{n}.$$

From the foregoing, one can see that the average salary of the group was successfully calculated, without asking group members for their salaries. Importantly, the privacy of the group was preserved while providing the answer sought. In fact, the way the problem was formulated did away with the privacy-transparency trade-off. The same approach can be applied in computing the average age of the group.

The lessons learned from the above can be applied to other system design problems:

- The way a problem is framed determines the number of constraints and the number of trade-offs.
- Reframing the problem can reduce the number of constraints and the number of trade-offs.
- Reducing the number of constraints and the number of trade-offs reduces problem/system complexity and ultimately simplifies system architecture.

In his book *Principles of Programming Languages: Design, Evaluation and Implementation*, MacLennan [29] borrows extensively from the teaching methods in architecture and aircraft design. He found that these analogies can be quite useful and serve as "intuition pumps" to enhance creativity. He found that several insights gleaned from architecture and aircraft design actually applied to programming language design.

8.4 Elegant System Designers

Designers of elegant systems possess a unique set of skills. Jeff Hawkins' fascination with cognition and neurobiology led to the invention of the PalmPilot. Art and history have been found to be especially inspiring for innovators. Einstein played Bach on his violin while he continued to delve into the mysteries of the universe. Simon [30] emphasized the exploration of the arts and suggested that creating complex designs is like working on an oil painting. He advocated greater interaction between engineers and artisans. He imagined engineers and composers carrying on a conversation about design and, in the course of that conversation, beginning to appreciate the common creative activity in which they both engaged. This recognition, he felt, would lead to their sharing experiences on the creative aspects of design. The Razor Pilot pen was inspired by the porous lotus root. Gutenberg's invention of the printing press, allegedly a combination of a wine press and a letter seal, was inspired at a wine festival. In every instance, these individuals reached beyond their respective disciplines for inspiration and, in so doing, came up with elegant solutions.

Steve Jobs, founder of Apple Computer, epitomized an elegant systems designer. Neither a computer scientist nor an industrial engineer, Jobs had a unique understanding

Table 8.3 Elegant designer characteristics

• Envisions the future with a variety of possible "what-if" assumptions
• Reasons analogically from other domains (e.g., biology, sociology, economics)
• Generates creative options by explicitly relaxing institutional, technical, economic, and sociocultural constraints
• Thinks critically and associatively (i.e., laterally)
• Engages in systems thinking [27] and design thinking [1, 2]
• Understands, exploits, and shapes user psychology
• Brings together people from different disciplines
• Stimulates collaborative thinking that fosters fluidity and serendipity
• Continually challenges conventions and norms

of humans and an uncanny ability to challenge norms as exemplified by Apple's tag line "think different." While we await the next Steve Jobs, it is certainly possible to identify the talents and skills needed to be an elegant designer (Table 8.3).

8.5 Smart Questions, Metaphors, and Analogies

An elegant design is ultimately the result of asking the right questions at the right time with context-awareness. Nadler's *Smart Questions* [31] offers key insights into how to go about asking the right questions. The key elements of this approach are people involvement (who should be a part of the solution creation effort), purposes (what are the potentially possible purposes for solving a problem to determine which purpose is really needed), future solution (what "ideal solution" should inform what to do in the present), and living solution (what can be done in the present and the future to achieve future solution state). To foster and perpetuate creativity throughout the design life cycle, the concept of divergence-convergence can be exploited. Specifically, this concept is operationalized through enumerating as many creative ideas as possible, organizing them into viable options that are responsive to the questions, and deciding on the right option for each specific phase in the system life cycle. Madni et al. [32] and Madni [33] offered a theory of creative option generation that employs psychological principles in conjunction with systems thinking to stimulate creativity and novel option generation – the precursors to elegant design. Engaging in such creative lines of inquiry can lead to new ways of thinking that increase the likelihood of realizing elegant solutions.

The right metaphor or analogy offers a powerful means to uncover an elegant solution. For example, 3D chips exploit a powerful everyday efficiency metaphor to deliver superior performance with lower power consumption [34]. This efficiency metaphor is best described as follows. Instead of driving to a different address in a suburb to get things done, one could employ an elevator in a skyscraper to accomplish the same objective. Clearly, elevators are more efficient than driving to destinations. The semiconductor industry has exploited this metaphor and analogy in 3D chip design. Instead of placing dies in separate packages, soldering to a circuit board, and sending data through their I/O points to other chips (i.e., driving through the suburbs), dies are stacked, and data is moved from one layer to the next (via the "elevator"). According to a semiconductor technology startup, a 3D design using two stacked dies with 22 nanometer geometry would produce much the same result (including reduced wire length, gate size, and device power consumption) as moving to one die with 15 nanometer geometry. According to Intel, the gates in a 22 nanometer transition are so small that over 4000 could fit across the width of a single human hair.

Another example of an elegant solution is Contour Crafting, an innovative, layer additive, fabrication technology [35]. Contour Crafting employs computer control and superior surface-forming capability of dual trowels to create smooth and accurate planar and free-form surfaces. Layered fabrication technology, which has been around for decades, builds 3D objects by stacking 2D slices. However, the application of this method is linked to relatively small objects because this approach cannot use thick layers without compromising surface quality. Contour Crafting's elegant solution is to build thick layers by placing an age-old trowel next to an extrusion nozzle. As the material is extruded, the outer surface of the layer is smoothed by the trowel under computer control. This approach does away with the limitation of existing approaches that are unable to maintain surface quality when attempting to build large objects. According to its inventor, an entire house or a colony of houses can be automatically constructed with required conduits for electrical, plumbing, and air conditioning in a single run. Some of the key advantages of Contour Crafting over competing layered fabrication processes are superior surface quality, higher fabrication speed, wide choice of materials, and lower cost. This technology has immediate application for ultrafast, low-income housing development and emergency shelter construction.

8.6 Heuristics-Enabled Elegant Design

People responsible for introducing elegance in system design are invariably system architects and product designers. In the absence of quantitative methods and analytical tools to assess design elegance, system architects and designers rely on intuition, experience, and judgment to develop elegant solutions to complex system problems. This approach is not without historical precedent. Prior to the development of the theory of strength of materials, civil engineers designed and built structures using heuristics or "rules of thumb." These heuristics, informed by historical successes and failures, were passed down through generations until, finally, the theory of strength of materials was born which offered a formal approach. Informed by this precedent, advances are required in several key areas. First, the academic community needs to engage in cross-disciplinary research that discourages the formation of discipline-specific academic silos. Second, the interaction between disciplines needs to be treated with the same rigor that is currently applied within disciplines. Third, the current compartmentalization of tools within engineering organizations is antithetical to elegant designs which requires a blurring of

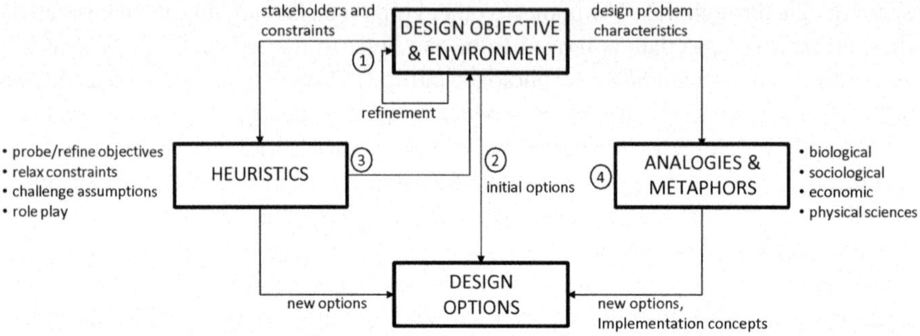

Fig. 8.1 Heuristics-enabled design process

boundaries between disciplines by making room for fluidity and serendipity during collaboration. Fourth, the engineering and design communities need to cease thinking about humans as imperfect processors of information, who need to be compensated for or shored up [36]. They need to find innovative ways of integrating humans with systems that capitalize on human creativity and ingenuity while circumventing human limitations [36]. Fifth, opportunistic use of appropriate analogies and metaphors from biology, psychology, sociology, and philosophy need to be exploited to achieve design elegance. Finally, metrics and heuristics (e.g., Ockham's Razor) for assessing design elegance need to be defined and collected during system design to inform and guide choices.

Against the foregoing backdrop, it becomes possible to define an interdisciplinary process for creating elegant designs. To begin with, this process needs to be iterative and initially heuristics-driven. It needs to foster systems thinking [27] and design thinking [36–38]. It needs to encourage associative thinking to explore and exploit metaphors and analogies from other disciplines and domains. In keeping with these tenets, Fig. 8.1 presents a heuristics-driven interdisciplinary process for elegant system design.

As shown in Fig. 8.1, the process begins by defining system objectives. Then, stakeholders and initial set of constraints are identified. These include operational, functional, physical, financial, regulatory, and social constraints. The initial option set is generated from predefined options (in the options library) associated with the design objective. Then, a variety of heuristics are applied to expand the design option space. These heuristics include (a) encourage users to identify higher-level objectives to prevent possible confusion between objectives and options, (b) identify known constraints and systematically surface implicit constraints by challenging assumptions, (c) incrementally relax constraints to open up the design options space, and (d) role-play various stakeholders (e.g., customer, user, maintainer, tester) to view design problem in a new light. The design problem characteristics are used to actively seek and exploit appropriate analogies and metaphors to generate new options and facilitate and simplify systems architecting and implementation. Thereafter, the candidate design options are evaluated using proxy measures of elegance (i.e., effectiveness, robustness, efficiency, predictability, and implementation ease). The proxies for effectiveness are usability/operability and perceived utility.

The proxies for robustness are physical and functional redundancy. The proxies for efficiency are number of steps to complete task and resource utilization. The proxy for predictability is reachability in the state space. The proxy for implementation ease is percentage reuse. The design option that gives the highest weighted score for elegance across all attributes is selected for implementation.

8.7 Assessment of System Elegance

Elegant system design can be viewed from multiple perspectives that collectively contribute to overall system elegance. These perspectives are overall experience, look, feel, behavior, and evolution. Associated with these perspectives are qualitative and quantitative metrics and measures of elegance.

Products, systems, and solutions to vexing problems can be deemed elegant if they offer an innovative combination of functionalities, or offer new functionality, or offer an innovative way of providing an existing functionality without having to engage in avoidable trade-offs and compromises. Sony's Walkman and Apple's iPhone are examples of elegant products. Sony's Walkman made music conveniently portable so one could jog and listen to music at the same time. Apple's iPhone offered an innovative, miniaturized combination of a computer and phone that allowed users to perform a wide variety of tasks that typically required several devices.

A compelling example of an elegant engineering solution is compressed sensing [39–42]. The value proposition of compressed sensing becomes readily apparent in the context of the Shannon-Nyquist sampling theorem usage. The Shannon-Nyquist constraint applies only when signals are encoded as a stream of evenly spaced samples. With other forms of encoding, the representation can be much more compact. In other words, the key insight is that signals of human interest have a sparse representation in some domain. This recognition leads to a general strategy for compression: (a) find a domain in which the signal of interest is sparse; (b) measure N ($0(10^6)$) components in this domain; (c) rank order the components from largest to smallest; and (d) retain the top k ($0(10^3)$) components and discard the rest. The challenge is finding the k largest values without making N measurements. The process is aided by the fact that the signal vector is sparse, i.e., only k of the elements is nonzero. This implies determining the positions and values of the nonzero elements with as few measurements as possible. This is precisely what compressed sensing does. Compressed sensing is an elegant signal processing solution with wide applicability. The ideas underlying compressed sensing came to Emmanuel J. Candès, a Caltech mathematician, when he was working on a problem in magnetic resonance imagery. Candès was pleasantly surprised that he could reconstruct a test image precisely despite the available data being insufficient according to the Nyquist-Shannon criteria. Working with Justin Romberg, he showed that the result was no accident and developed much of the theory underlying compressed sensing. Thereafter, their collaboration with Terrence Tao of UCLA produced the principles of compressed sensing. They demonstrated that the efficiency of their method was close to the theoretical optimum. The reason compressed sensing is an elegant

Table 8.4 Elegant system metrics

Characteristic	Metric
Simplicity	
Purposivity	Goal achievement level; number of side effects; number of unintended consequences; cycle time
Parsimony	Total number of component elements; resources used; interventions made; structural complexity; behavioral complexity
Transparency	Observability/deducibility (without assumptions) of state of system execution and status of components
Usability	Time to mistake-free usage; learning time; cognitive complexity
Power	
Scalability	Percent increase in implementation/manufacturing cost with increase in number of elements and interconnections
Sustainability	Ability to adapt to changes in the environment (competition, regulation, technology)
Bonding	User base size/growth; turnover in user base (loyalty)
Efficiency	Cycle time; resource utilization
Evolveability	Time to evolve; cost to evolve
Relevance	
Affordability	Total costs; design costs; manufacturing costs; maintenance costs
Utility/Impact	ROI; lives saved; increase in quality of life; perceived worth
Predictability	Ability to foretell system behavior in different circumstances

solution is because it challenged conventional thinking that people tend to erroneously associate with the Nyquist-Shannon theorem. By interpreting this theorem correctly, compressed sensing exploits principles of *sparsity* and *incoherence* to design efficient sensing (or sampling) protocols that capture and condense useful information contained in a sparse signal, into a small amount of data. What is remarkable about this approach is that it allows a sensor to efficiently capture the information in a sparse signal without trying to comprehend the signal. Today compressive sensing is used in a variety of generic applications such as data compression, channel coding, and data acquisition.

The foregoing discussion provides the basis for specifying qualitative and quantitative measures of system elegance. Table 8.4 identifies and defines key metrics that collectively define system elegance. *Simplicity* stems from purposivity, parsimony, transparency, and usability. *Power* stems from scalability, sustainability, bonding, efficiency, and evolvability. *Relevance* stems from affordability, utility, and predictability. Simplicity, power, and relevance are the cornerstones of elegance.

8.8 Summary

Elegant systems design has become an area of interest in both academia and industry in the last decade. However, the literature on elegant design is sparse and fragmented. For the term "elegant design" to apply, the system design problem has to be necessarily complex. For complex systems, an elegant solution exhibits economy of means to conquer complexity and resolve/dissolve complications to produce economy of form. In simple terms, elegance is an effortless display of sophistication. Elegant design can shape our future by creating cultural values which, in turn, influence our future. As important, an elegant design creates and fosters a relationship with the user or audience. Elegant design embodies a few fundamental truths: sophistication need not imply complexity; designs should be self-explanatory; and complexity can often be reduced through proper problem reframing, and/or use of appropriate analogies/metaphors.

This chapter formalizes the characteristics of elegant systems and consolidates the different perspectives associated with elegant systems design. These include stakeholder needs, elegant system designer characteristics, and various approaches to creating elegant systems. Based on these findings, a heuristics-driven process for elegant system design is presented along with metrics for assessing elegance. Collectively, these perspectives and overall process are intended to allow systems architects and product designers to understand and apply the principles of elegant systems design.

In today's environment, in which affordability, manufacturability, cyber and physical security, adaptability, and resilience are viewed as the most important quality attributes, the pursuit of elegance in system design has not received the attention it deserves. And yet, it is elegant systems that capture our imagination and that create that all-important emotional connection. It is my hope that this chapter will inspire others to pursue research in this nascent area with a view to developing intelligent design environments that foster the creation of elegant systems.

References

1. Cropley D, Cropley A. Engineering creativity: a systems concept of functional creativity. Creativity across domains: Faces of the muse. 2005:169–85.
2. Gelernter DH. Machine beauty: elegance and the heart of technology: basic books; 1998.
3. O'Connor J. Elegant design for everyday life. Harv Bus Rev. 1990:134–9.
4. Billow SA. The role of elegance in system architecture and design, S.M. Thesis, Massachusetts Institute of Technology; 1999.
5. Lewis K. Making sense of elegant complexity in design. J Mech Des. 2012;134(12):120801.
6. Simons CL, Parmee IC. Elegant object-oriented software design via interactive, evolutionary computation. IEEE Transactions on Systems, Man, and Cybernetics, Part C (Applications and Reviews). 2012;42(6):1797–805.
7. Morgan M. Vitruvius Pollio "The ten books on architecture.". New York: Dover; 1960.
8. Edwards O. Elegant solutions. New York: Three Rivers Press; 1989.
9. Rubinstein MF. Patterns of problem solving. Englewood Cliffs: Prentice-Hall; 1974.
10. Gell-Mann M. The quark and the jaguar: adventures in the simple and the complex. New York: St. Martins Griffin, Macmillan; 1995.
11. Gernert D. Ockham's razor and its improper use. Cognitive Systems. 2009;7(2):133–8.
12. Rechtin E. Systems architecting: creating & building complex systems. Prentice Hall: New Jersey; 1991.
13. Eppinger SD, Ulrich KT. Product design and development 1995. 1995.

14. Maier MW. The art of systems architecting: CRC Press; 2009.
15. Deck M. The power of product platforms: building value and cost leadership by Marc H. Meyer and Alvin P. Lehnerd. New York : The Free Press; 1997. 267 + xiv pages. $dollar;35.00. J Prod Innov Manag 1997;14(6):526–9.
16. Brooks FP. The mythical man-month. 2010.
17. Bellman R. Personal communications with colleagues at the University of Southern California, Los Angeles 1971.
18. Kalaba R. Personal communication with professor Firdaus Udwadia of the University of Southern California, Los Angeles 1975.
19. Warwick G, Norris G. Is it time to revamp systems engineering. Aviation Week. 2011.
20. Griffin MD. How do we fix system engineering. 61st annual international congress. Prague; 2010.
21. Madni AM. Transdisciplinarity: reaching beyond disciplines to find connections. J Integr Des Process Sci. 2007;11(1):1–11.
22. Walker R. The guts of a new machine. The New York Times. 2003; 30.
23. Lakshminarayanan V, Liu W, Chen CL, Easterbrook S, Perry DE. Software architects in practice. Report (Empirical Software Engineering Lab (ESEL), University of Texas, Austin, USA, 2005). 2005.
24. Discovery GPSD. On understanding, learning, and teaching problem solving. New York: Wiley; 1957.
25. May ME. The elegant solution: Toyota's formula for mastering innovatio. New York: Simon and Schuster; 2007.
26. Ackoff RL. Creating the corporate future: plan or be planned for. Hoboken: Wiley; 1981.
27. Senge PM. The fifth discipline: the art and practice of the learning organization. New York: Broadway Business; 2006.
28. Leff RD. Elegant solutions. Cambridge: Massachusetts Institute of Technology; 1997.
29. MacLennan BJ. "Who cares about elegance?" The role of aesthetics in programming language design. 1997.
30. Simon HA. The sciences of the artificial. Cambridge: MIT Press; 1996.
31. Nadler G, Chandon W. Smart questions: learn to ask the right questions for powerful results: San Francisco, Wiley; 2004.
32. Madni AM, Brenner M, Costea I, MacGregor D, Meshkinpour F, editors. Option generation: problems, principles, and computer-based aiding. Proceedings of the 1985 International Conference on Systems, Man, and Cybernetics, Tuscon; 1985.
33. Madni AM. Generating novel options during systems architecting: psychological principles, systems thinking, and computer-based aiding. Syst Eng. 2013;17(1):1–9.
34. Wood L. 3D chips: the next electronic revolution computerworld. 2011.
35. Khoshnevis B, Russell BR, Kwon H, Bukkapatnam S. Contourcrafting—a layered fabrication technique. IEEE Robot Automat. 2001.
36. Madni AM. Integrating humans with software and systems: technical challenges and a research agenda. Syst Eng. 2010;13(3):232–45.
37. Thomke SH, Feinberg B. Design thinking and innovation at Apple. Cambridge: Harvard Business School; 2009.
38. Six JM. Design Thinking| Employing Design Principles| Defining Ease of Use. ASK UXmatters, Jan 17, 2011.
39. Candès EJ, Romberg J, Tao T. Robust uncertainty principles: exact signal reconstruction from highly incomplete frequency information. IEEE Trans Inf Theory. 2006;52(2):489–509.
40. Baraniuk RG. Compressive sensing. IEEE Signal Process Mag. 2007;24(4):118–21.
41. Candès EJ, Wakin MB. An introduction to compressive sampling. IEEE Signal Process Mag. 2008;25(2):21–30.
42. Hayes B. The Best Bits American Science. 2009:276–80.

Chapter 9
Affordable Resilience

"One should expect that the expected can be prevented, but the unexpected should have been expected."

– Norm Augustine, national leader in science, technology, business executive

9.1 Background

Cost-effective protection of complex systems and infrastructures from failures and performance degradation caused by disruptions such as physical security and cyber-security breaches has been a systems engineering design goal and a national imperative for well over a decade [1, 2]. This capability falls under the rubric of resilient systems. Resilience means different things in different domains (e.g., military, space, healthcare, energy). For example, in the military domain, resilience is defined as the ability of a system to adapt affordably and perform effectively across a wide range of operational contexts, where context is defined by mission, environment, threat, and force disposition [1]. A key issue in engineering resilient systems is the lengthy and costly upfront engineering process. As important, current approaches to resilient system design rely on ad hoc methods (e.g., safety nets) and piecemeal solutions when developing mechanisms to respond to disruptions and unanticipated system behaviors [3, 4]. In such approaches, *observed* high-level behaviors are compared to *expected* high-level behaviors to determine when the difference exceeds an acceptance threshold. When it does, the observed behavior is considered a problem or a precursor to a problem. Such behaviors trigger a transition to a known safe state until the underlying problem is diagnosed and resolved. During the problem resolution, the system remains unusable. Furthermore, existing methods do not take into account the different states and modes of complex systems, nor do they address unprecedented disruptions that can occur at arbitrary times during complex system operation. They also do not address the time-dependent nature of disruptions and their impact on complex systems. In light of the foregoing, there is a pressing need for a resilient systems development methodology. This recognition provides the motivation for exploiting formal modeling approaches for engineering resilient systems. The formal modeling approach needs to have sufficient flexibility in its formalisms to accommodate uncertainty in system states that result from partial

A.M. Madni, *Transdisciplinary Systems Engineering*,
DOI 10.1007/978-3-319-62184-5_9

observability of system behavior. The models should also be amenable to verification, testing, and learning from observation. To advance beyond the state of the art in model-based approaches requires the ability to determine appropriate desired behaviors of complex systems. This is a key challenge because complex systems invariably have a large state space, with some "hidden" states arising from complex interactions between system elements and between the system and the environment. Exacerbating the problem is the fact that the state of the system is often not known because of partial observability of the system and environmental uncertainties. Additional complicating factors include incomplete understanding of system dependencies and environmental influences, likelihood of conflicts between local and global responses, and increase in the number of human roles.

9.2 DOD Perspective on Resilience

With increasing system complexity, it is becoming increasingly apparent within DOD establishments that engineering design and development processes need to be transformed with the help of promising, new technologies. The transformed processes and attendant tools need to be able to: (a) span the system life cycle from concept formation through sustainment; and (b) support both rapid fielding activities as well as traditional acquisitions.

In DOD and military terms, resilience is defined as robustness that is achieved through thoughtful, informed design that makes systems both effective and reliable in a wide range of contexts and readily adaptable to others through reconfiguration or replacement. A resilient DOD system also displays detectable, graceful degradation of function when pushed outside its base capabilities and operating envelope. The engineering challenge of the twenty-first century is to incorporate resilience into systems using a disciplined approach and in an affordable, timely manner [5].

This view of robustness is in sharp contrast to the historical approach for achieving robustness, which has been through overdesign, a clearly unaffordable strategy. At the same time, it is unaffordable to procure systems that meet performance specifications through ad hoc interactions among highly interdependent and tightly coupled subsystems because upgrading/configuring such systems to operate in a new operational environment is not only cost prohibitive but also unsustainable in the long run. While such designs are occasionally a consequence of cutting corners in engineering, at other times they are the result of lack of anticipation. By now, the systems engineering community has come to realize that it is not possible to satisfy competing demands to be better, cheaper, and faster by cutting corners. As important, the new metrics for DOD systems in the 21st century are affordability, adaptability, and resilience. Thus, the ability to analyze system elements and their interactions in sufficient depth is critical to making informed decisions.

In light of today's geopolitical uncertainties, DOD systems are called upon to perform increasingly more complex missions in a variety of operational environments. Therefore, these systems need to be rapidly deployable, affordable, adaptable, and highly reliable [6, 7]. These characteristics collectively define *resilient* systems. In particular, the *affordable and adaptable* characteristics imply design for adaptability as well as cost-effective reconfiguration and replacement. The latter is also a key requirement for system of systems (SoS). From a DOD perspective, the key issues are whether or not a system or SoS can be made affordable, adaptable, and effective. If not, can reconfiguration or replacement achieve these ends? A related issue is whether or not reconfiguration and replacement can be performed better and faster than is possible today. Fortunately, with advances in model-based engineering (MBE), platform-based engineering (PBE), parallel computation, and data analytics, it is becoming possible to design smarter, avoid costly and time-consuming rework cycles, and compress overall cycle times [8]. To achieve these goals (i.e., end state) begins with understanding how design is accomplished today.

Today substantial effort is being expended on design. The percentage of engineering work relative to program cost in DOD has risen from 10% to 20% in the 1950s to 40–60% today. Consequently, there has been a steady erosion in the time and money that can be spent on purchasing components/subsystems. The continuation of this trend implies loss of economies of scale, reduction in quality, and increased rework. It is not just because systems are much more complicated today; it is because current methods, processes, and tools have failed to keep pace with increasingly complex interactions among components and disciplines, as well as with increasing operational demands. As important, engineering productivity is impeded because less upfront engineering is being done today. As a result, problems are being discovered late in the development cycle making them expensive to fix. Often, the only viable option is to discard prior work and start over. Meanwhile, the DOD customer has no recourse but to patiently wait. Finally, this approach does not address system affordability or the price of the product, nor does it address the trade-offs that need to be made to achieve cheaper, better, and faster results [9]. According to Kendall [10], affordability results from reducing engineering cycles and buying COTS, whenever possible.

9.3 Affordability Imperative

In the world of systems acquisition more so than other activities, time is money. The majority of costs stem from amortizing lengthy development and test cycles over relatively small quantity purchases [7]. And, of course, when engineering issues arise, causing delays in a development program, the "cost meter" continues to run driving up costs and potentially reducing the quantity that can be purchased. Eventually, the amount of rework becomes unaffordable invariably leading to program cancellation. To make engineering development affordable requires up front collaboration among all stakeholders. The latter is the key to reducing the risk of schedule slips and cost overruns on acquisition programs.

Affordability requires advances on several fronts: accelerating engineering processes, circumventing or reducing the time spent on test-and-fix, doing more extensive engineering design and testing faster, doing more in parallel, and using greater computational power to enhance coverage of the preceding activities. Subsequently, decisions about how best to invest the savings can be made (e.g., buy or procure additional systems). To effectively accomplish the above requires *significant communications among and across engineering and acquisition activities* to understand scope, identify interdependencies (that can potentially lead to undesirable outcomes), and make informed trade-offs and decisions. To accomplish these objectives, especially for adaptable systems, it is imperative to understand and explore the range and likelihood of potential situations (defined by factors such as mission, environment, threat, and concept of operations). It is also crucial for generating a range of relevant use cases and test scenarios. While system adaptability is clearly desirable, *the testing and evaluation to ensure design success* is just as important, if not more. Historically, testing and evaluation address traditional concerns about performance and reliability. In this regard, risk-oriented and uncertainty-oriented techniques such as real options [11–14] and importance sampling [15] can be potentially effective in reducing the cost of testing and evaluation. An additional evaluation concern is the trade-off decision that balances adaptability against performance, reliability, and other measures.

Affordability, adaptability, and effectiveness are inextricably linked in systems engineering. A resilient system operates within the area defined by an effective balance among these attributes and satisfies customer needs and circumstances. As importantly, a resilient system maintains this balance throughout its lifetime.

9.4 A Common Misconception

A common misconception in some engineering circles is that the engineering of resilient systems can be entirely addressed as a process problem, rather than as a science and technology problem [1]. In fact, some believe that "if only" current processes were rigorously followed, the problem would take care of itself. Others would rather reengineer using well-known techniques [16–19]. To add perspective to this discussion, it is worth recalling that from the days of Robert McNamara through the Packard Commission and beyond, engineering and acquisition processes have undergone continual improvement and refinement, coupled with diligent efforts to enforce compliance. The results have been less than promising [20, 21]. Despite fifty years of process reforms, engineering and acquisition processes have not succeeded in controlling costs, adhering to schedules, or satisfying performance requirements. The main reasons for this lack of success are prematurely culling alternatives, making decisions without adequate information, employing processes that are sequential and slow, losing information at every step, and continuing to refine requirements in an ad hoc fashion.

Systems engineering studies repeatedly show that problems discovered late in the system development life cycle can be up to a hundred times more time-consuming and expensive to fix [7, 22]. Furthermore, the less upfront engineering is done, the more likely it is that the program will fail. Even so, for program managers to invest in the necessary upfront engineering, it is imperative to significantly reduce the cycle time and cost of upfront engineering. In the absence of needed upfront engineering, the only recourse is to resort to process reforms that are wholly dependent on front-loading program costs. In today's environment, this approach is hardly viable. Fifty years of combating the misperception that upfront engineering is nothing more than an expensive nuisance should have taught the systems engineering community an important lesson: upfront engineering needs to be much cheaper, faster, and collaborative than it is today. In fact, upfront engineering should serve as a "forcing function" to accelerate problem solving, not merely a means to discover design flaws and defects faster.

To understand the relevant science and technology issues, it is important to get past the common misconception that systems engineering is merely about adhering to process "recipes" and working from handbooks of standard practices. By unfairly characterizing systems engineering as a discipline that tells us what we can and cannot do, its real contributions are unfortunately overlooked. Systems engineering is about *identifying interactions* among and across component elements of the phenomena of interest, *understanding consequences* of those interactions, and *exploring ways to effectively manage* them.

However, this forward-looking perspective raises certain fundamental concerns. How does one describe and analyze devices, the environment, and the behavioral elements that create the interaction between them? What are the computational mechanisms to rapidly generate and test ideas? Given that even these mechanisms are not going to overcome the computational complexity associated with exhaustive analysis and testing, what are the techniques for assessing uncertainty and risk that can help maximize the payoff of investment in engineering work, minimize the likelihood of having to do extraneous work (i.e., work that we choose not to pay for), and help us decide how much to invest? These are some of the deep problems that make the engineering of resilient systems a science and technology challenge. Having a sound understanding of these issues is a prerequisite to developing effective and enduring solutions.

9.5 Technology Challenges and Needed Advances

Upon investigating how these fundamental problems and engineering activities relate to each other, it becomes apparent that part of the engineering challenge is greater demands on system design. Systems have to satisfy more constraints today than ever before. In the past, systems were designed to be *effective*; now they also need to be *affordable* and *adaptable*, with significantly fewer design-test-build cycles. To rapidly create and field affordable and effective systems, programs must conduct comprehensive systems engineering trade-off analyses to arrive at balanced solutions. Specifically, the solutions need to explore factors such as system adaptability, trust, and affordability in the system design trade space [1]. These considerations need to be revisited when modifications are made during design, manufacturing, and fielding. As important, engineers need requirements that are more informative than what they are provided to them today. Requirement

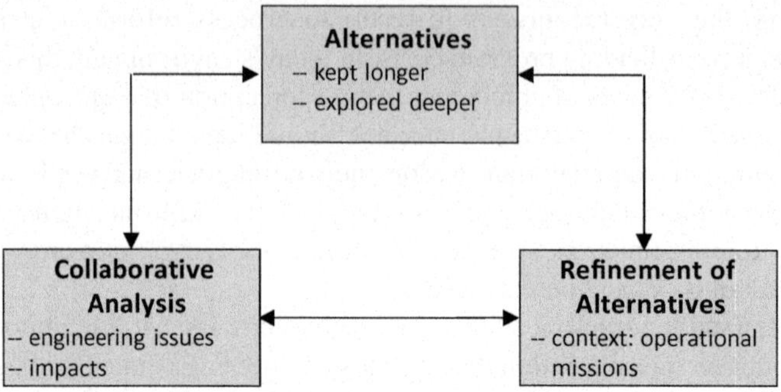

Fig. 9.1 Key interactions in identifications and refinement of alternatives

refinement needs better grounding in design feasibility and comprehensive exploration of opportunities. Finally, options need to be more thoroughly explored and kept open longer than they are today [23, 24]. Figure 9.1 illustrates the interrelationship among these considerations.

New methods and tools are needed to enable design for adaptability, effectiveness, and timeliness. In particular, there is a pressing need for models with the requisite semantics to represent different types of designs and to enable more detailed analysis of design properties than is possible today [25]. Also needed are tools that collect stakeholders' inputs as well as empirical information relating needs to human behavior. Effective designs result from analyzing what people actually do (i.e., their behavior), not just what they say they want (i.e., their desires), or what they say they do (i.e., their beliefs). Fortunately, access to ever-increasing, massive computing power allows deeper consideration of trade-offs and options. Used appropriately, this capability is an important defense against both tactical and technological surprise.

Model-based engineering (MBE) and platform-based engineering (PBE) are useful starting points for developing new engineering tools and environments [26, 27]. They enable the exploration of design alternatives and adaptability choices by offering computational means to evaluate intra-system characteristics and exploration of system interactions with the external environment. The challenges lie in formulating specifics. The following paragraphs present key challenge areas that matter most along with needed technological advances.

System Representation and Modeling Even though models have become a central aspect of systems engineering with the advent of MBSE, systems modeling is not synonymous with systems engineering. In reality, systems modeling and system models are enablers of systems engineering. They allow representation of multiple types and perspectives needed to capture both the physical and logical structures, as well as system behaviors, system interactions (with the environment), and system interoperability with other systems and systems of systems (SoS).

Upon closer examination of the different kinds of models required and the different disciplines, aspects, and phenomena they need to address, it becomes apparent that there is a pressing need to create and manage multiple classes (e.g., executable, depictional, statistical) and multiple types (e.g., device and environmental physics, communications, sensors, effectors, software, systems) of models. Taking the interactions among these models into account, it can be concluded that there are dozens of different models that need to be developed and made interoperable. Their form and content, and the rate at which they can be created and validated, are key to filling existing gaps. These gaps can be addressed with the creation of models and simulations combining live and virtual elements; the acquisition and cross-integration of physics-based and statistical models; the building and integration of critical multidisciplinary, multiscale physics models; automated and semi-automated techniques for acquiring models; and techniques and tools for building adaptable models.

Characterizing Changing Operational Environments A critical challenge today is how best to complement system models with models of the dynamic operational environment needed to drive system behavior [28]. Developing such models begins with a deep understanding of customer needs. Therefore, it encompasses gathering and modeling operational data and experimenting with alternative designs to understand the operational impacts of various alternatives.

Today, design and test activities are conducted to satisfy requirements, such as achieving desired performance on key metrics under certain conditions. However, this approach ignores what complexity theory and practical experience have taught us – optimizing in this fashion invariably leads to brittle systems. While such solutions may achieve high performance under certain conditions ("point-based design"), they fail to fully and adequately address the *range* of conditions likely to be encountered. To design and test for resilience across a broad range of conditions requires understanding at a much finer-grained level of how the systems are expected to be used (i.e., system concept of operations), the environments in which they will operate, the likely threats, and the operational context.

The US Army Tank Automotive Research, Development, and Engineering Center (TARDEC) Ground Vehicle Systems Engineering Team (GVSET) News noted that DOD spent a decade trying to get ground vehicles to be more fuel efficient with existing functionality [29]. TARDEC built a set of Fuel-Efficient Ground Vehicle Demonstrators to illustrate these points. In addition, there were instances of vehicle designs that were created to satisfy unrealistic requirements. For example, a Humvee today accepts an increase in weight while giving up fuel efficiency, in part, to meet the requirement of carrying four 6′5″ tall soldiers. An analysis conducted by a TARDEC contractor showed that the statistical likelihood of four men this tall being in the same unit is extremely low. Furthermore, fuel economy tests were conducted with these vehicles driving a fixed circuit over a fixed terrain. While these conditions were representative of many driving conditions, the tests conducted did not take into account the frequency with which military missions encountered those conditions. Not surprisingly, these vehicles performed well on tests but disappointed in actual use. This discrepancy has also been encountered with automobiles that were purchased solely on their EPA ratings. EPA ratings do not reflect the realities of actual vehicle usage. When TARDEC's contractors collected and statistically modeled data on actual driving conditions in the theater, they were able to design vehicles that performed significantly better both on tests and in operational environments.

These simple examples argue for the value of real-world data collection and data analytics in vehicle design. Real-world data collection in operational environments comes with exciting opportunities and challenges. For example, today both social computing systems and physical systems and environments are being increasingly instrumented for real-world data collection for a variety of purposes ranging from marketing to maintenance. These capabilities can be opportunistically leveraged in engineering design and development. In addition to exploiting instrumented systems and environments, a range of technologies such as augmented reality and virtual worlds are being increasingly used to collect data, present practice scenarios, and project future stress conditions for training and mission planning. Leveraging these advances for engineering can produce faster payback and delivery of interoperable systems to a variety of potential users (i.e., planners, operators, trainers) of engineering models of physical devices. In addition, the use of test data to refine computational models remains another underexploited area in that test data, currently collected, is not used for this purpose [6, 30].

In sum, the technologies needed to fill specific gaps in characterizing changing operational environments include *instrumentation* to collect data from live and virtual operational environments, systems, and system tests; *virtual worlds* for experimentation and learning; *automated and interactive acquisition* of operational context models (missions, environments, threats, tactics, and CONOPS); *abstraction and generalization* of tests and use cases from operational data; and *synthesis and use* of behavioral and environmental models within virtual worlds.

Jeffrey Holland, former Director of U.S. Army's Engineering Research and Development Center (ERDC), argued that, "ensuring adaptability and effectiveness requires evaluating and storing results *from many, many scenarios* (including those presently considered unlikely) *for consideration earlier* in the acquisition process." A decade from now, the combination of virtual worlds, augmented reality, and multi-visualization models will make it possible for the first time to assess the breadth of mission effectiveness of complex weapon systems in relevant operational contexts. This advance needs to begin with characterizing changing operational environments and using the resultant models informed by data analytics to "drive" system behavior.

Cross-Domain Coupling Several models discussed in the preceding paragraphs exist today, but very few are interoperable. These stand-alone models need to be made interoperable and possibly expanded. The end goal is to be able to model complex systems fully across multiple domains (e.g., materials, fluids, physics) and operational environments.

To make heterogeneous models and model classes and types work effectively together requires new computing techniques in addition to standards. While standards are clearly part of the solution and facilitate interoperability, they do not provide the actual solution. Models can differ in type, detail, coverage, representation, data requirements, and several other aspects (e.g., efficiency, maintainability, knowledge depth, availability of required skillset, costs, development phase, demand, and market structures and forces) that need to be included for good reasons. Thus, composing system models from heterogeneous components presents challenges that go beyond mere adherence to standards. The two most important challenges are accomplishing information interchange between incommensurate models and resolving temporal, multiscale, and multi-physics integration mismatches.

Several of these issues can be resolved by creating libraries with reusable content, developing capability to compose models rapidly and correctly, accelerating workflow definition between models, and automating conversion between specific models. To realize these capabilities, research is needed in on-demand composition of workflow for modeling and analysis; consistency maintenance across hybrid models through data abstraction and spatiotemporal data exchange; efficient interoperation through automatic generation of summaries and "surrogate" models; creation and repair of mappings between modeling systems using semantic features; and interface extension programming to provide automated boundary condition assignment (parameterization), coordinated cross-phenomena simulations, interfacing to decision support, and coupling and connections to virtual worlds and augmented reality.

Trade-Space Analysis The prior discussion has been confined to the core capabilities needed for rapid and sophisticated analysis and testing. However, to fully realize the needed advances requires extending the ability to identify and understand problems and to more rapidly develop effective solutions. Doing so requires the ability to rapidly generate a substantially larger set of alternatives than is possible today and to understand the implications of choosing one alternative over the others. While the required capabilities and tools described up to this point will certainly enhance our ability to do so, more can be accomplished with the ubiquity of computing and computing advances. These create opportunities for generating more options than previously possible, exploring them in greater detail, and keeping them open longer while assuring that the resulting complexity can be managed [23, 24]. Computing advances also enable more extensive testing.

To enhance productivity requires tools and "drivers" to efficiently generate and evaluate alternative designs. Specifically, there is a need to automate the exploration of a large number of conditions, generate and test more alternative solutions, analyze the resulting data, and rapidly deliver the findings and relevant trade-offs to decision makers. As important, the ability to rapidly evaluate available options in multidimensional trade spaces is an important consideration. Thus, a key challenge is that of control, i.e., engineers need to be able to target these drivers effectively because despite dramatic advances in computing, the available computational power is still inadequate to explore every possibility.

The advances needed in this area to fill existing gaps include guided automated searches and context-sensitive search algorithms, ubiquitous computing for generating and evaluating options, identification of high-impact variables and their likely interactions, new sensitivity localization algorithms, algorithms for measuring adaptability, risk-benefit and cost-benefit analyses tools, integration of reliability and cost into acquisition decisions, and cost-sensitive and time-sensitive uncertainty management via experimental design and activity planning. The massive trade spaces associated with complex systems [31] make filling these gaps a key imperative.

Collaborative Design and Decision Support Ultimately, all technological challenges involve or can be traced back to people [1, 6, 32]. The people challenge encompasses both providing information to and acquiring information from people. The technologies needed to facilitate information exchange in system design include distributed collaboration, information summarization and abstraction, multimedia presentation, and human computer interaction.

For reasons provided under characterizing changing operational environments, information needs to be acquired from a much wider range of stakeholders than is done today. It is tragic, for example, that service members returning from deployments routinely complain about their equipment but without knowing who to contact for problem resolution. Today, the time between when soldiers first start throwing sandbags into the bottom of their vehicles and when engineers first discover an underbelly blast problem and begin offering viable alternatives to soldiers needs to significantly be shortened. Also, much more complex information needs to be conveyed to decision makers than is done today because decision makers today are offered only opinions and anecdotes as inputs for critical decisions made over a limited set of alternatives. In contrast, the engineering of resilient systems entails making empirical data-driven decisions about a rich set of critical alternatives. In so doing, non-engineers need assistance to make realistic engineering assessments of feasibility of various options and opportunities. The foregoing does not imply that every individual will be communicating with every other about everything, all the time. Rather, the challenge is to create an environment that supports context-driven, focused information exchanges.

The key gaps that exist today in collaborative design and decision support can be filled with the creation of usable, multidimensional trade spaces; decision rationale capture; trade-off prioritization aids; explainable decisions; engineering, system acquisition, physics-based, and behavioral models amenable to V&V; access controls; and effective information push-pull that does not exceed cognitive capacity of humans [6, 33].

9.6 Promise of Formal Methods for Resilience Engineering

The safeguarding and protection of engineered systems from failures and disruptive events is a systems engineering design goal and a national imperative. Even so, progress has been slow resulting in the use of traditional fault-avoidance and fault-tolerance methods in conjunction with ad hoc, piecemeal resilience approaches. However, fault-tolerance methods are “inward-looking,” i.e., they focus on the system and its components, not external disruptions (e.g., human error, system misuse, and unanticipated conditions in the operational environment). They tend to be difficult to verify, and do not generalize, or scale. Explicit representation techniques such as finite state machines are inadequate because they suffer from combinatorial explosion and, therefore, do not scale. On the other hand, generative AI planners are not amenable to V&V, an essential requirement for reliable and predictable operation. This recognition provides the impetus for developing formal methods that enable verification and validation and generalization, and that scale with increase in system complexity. This capability is needed by both the DOD (e.g., control of distributed heterogeneous UAV swarms) and commercial sectors such as automotive (e.g., self-driving autonomous SoS), energy (e.g., adaptive grids), and healthcare (e.g., patient surge handling).

Formal methods are conspicuously absent in resilience engineering for two main reasons. First, lack of research in this area has resulted in people relying on ad hoc methods such as safety nets that came out of the world of fault tolerance. Second, the definition of the term "resilience" itself tends to be wide ranging and largely context dependent. Therefore, it is not uncommon to find the word "resilience" being used in a way that is synonymous with fault-tolerance or robust design (robust control theory). To prevent ambiguity in definition, I define resilience as the ability to recover fully/partially from the negative effects of a disruption within a specified time period, dynamically extend capacity and resources to respond to a disruption of a certain scale, and restructure or reconfigure the system to minimize the impact of disruptions.

Resilience has been addressed in a wide variety of contexts ranging from adaptive organizations [34–37], psychology [38, 39], safety-critical systems [40, 41], and ship design [42]. Resilience means different things in different domains and to different people [30, 40, 43–49]. Engineered resilience is a system property that allows a system to continue to provide useful service in the face of largely unpredictable, disruptive events which can be internal or external to the system. The types of disruptions fall into three categories: external disruption – caused by factors outside the control of the system; systemic disruption – service interruption due to an internal fault; and human-triggered disruption – the result of human error or system misuse [44]. System behaviors that exhibit resilience include the capability to:

- *Circumvent disruption* – this means the system has the ability to anticipate and avoid having to confront the disruptive event; this capability has been addressed in the body of work called fault avoidance and obstacle avoidance.
- *Withstand disruption* – this means being able to confront and endure the disruption without degradation in performance within the system's performance envelope; this capability has been addressed in the body of work called robust control.
- *Recover from negative effects of a disruption* – this means the system suffers temporary degradation in performance but is able to recover from the disrupting event to an acceptable degree and within an acceptable duration; that both the disruptive event and system response can potentially leave both the system and environment changed from their pre-disruption states.

- *Dynamically extend capacity* – this means adding resources on demand to counter the disruption; the added resources may be released post-disruption or repurposed.
- *Restructure/reconfigure system* – this means altering the participating components (or nodes in an SoS), as well as linkages and information flows among them to minimize/mitigate the impact of the disruptive event to assure continuity of system service (e.g., cybersecurity breach).

With these definitions, system failure means that the system is incapable of providing useful service. Of the five types of resilient behaviors presented above, the first two are covered by existing methods from fault avoidance and robust control theory. The remaining three require advances in theory, concepts, and new methodologies. These three, which require real-time tradeoffs, fall under the rubric of engineered resilience. For example, to recover from negative effects of a disruption, there is a trade-off between the degree of recovery and the time to recovery. Similarly, to dynamically extend capacity there is a trade-off between the amount of added capacity and the cost of adding that capacity. Finally, in restructuring or reconfiguring the system, there are trade-offs involving time to restructure/reconfigure, degree of restructuring/reconfiguring, the added complexity from the restructuring/reconfiguration, and the cost of restructuring/reconfiguration.

Formal methods comprise techniques for modeling complex systems in a mathematically rigorous fashion. The resultant models enable verification of the system's properties more thoroughly than would be possible using purely empirical testing. With formal methods, there is an important trade-off between the level of rigor and the degree of flexibility in the model to capture complex behaviors. For example, both system safety and resilience become important considerations. In this regard, a formal approach can effectively complement system testing to ensure correct behavior [50]. In contrast to traditional system design methods, formal methods employ formal verification schemes to ensure that the basic principles governing system behavior are proven correct before they are accepted. It is important to note that formal verification cannot and does not circumvent the need for traditional testing because formal verification cannot fix unwarranted (i.e., poor) assumptions in design. However, formal verification can help identify reasoning errors that would otherwise go unidentified and unresolved.

Formal methods can be deterministic or stochastic. The choice of modeling approach depends on factors such as observability, measurement uncertainty, and need to adapt to changes in the operational environment. The choice of deterministic or stochastic methods is not an either-or proposition. The two approaches can co-exist and, in fact, can be combined to exploit the benefits of each. Examples of deterministic methods include computational tree logic, linear temporal logic, and contract-based design. Examples of stochastic methods are Partially Observable Markov Decision Processes (POMDP) and Hidden Markov Models (HMM). These are black box models in the sense that the underlying states, state transitions, and state emissions are either unknown or only partially known at the outset. However, these models can be "trained" to recognize nominal behavior and flag unusual behavior through the use of appropriate reinforcement learning techniques.

Formal methods offer a rigorous approach to specifying system requirements. In particular, *assume-guarantee* construct that underlies contract-based design (CBD) is especially relevant. CBD specifies guarantees on system behaviors under specific assumptions about the environment. CBD is a compositional approach to design that reduces complexity in design, implementation, and verification by decomposing system-level operations (i.e., tasks) into manageable subproblems. In complex systems, specifications can be incomplete or have inaccuracies. Uncertainties about the environment can further complicate matters. Consequently, adaptability is crucial to ensure that the system continues to operate as intended in the face of uncertainties and disruptions. Formal modeling based on extending contract-based design can help with the design, implementation, and verification of adaptive systems. Traditional contracts that offer a formal framework for system specification, provide a suitable starting point for building flexible contracts that are needed for modeling adaptable/resilient systems.

The essential characteristics of a systems model that would inform the development of resilience mechanisms include explicit consideration of uncertainties and risks, nominal and predictable off-nominal behaviors, unexpected behaviors (acceptable, unacceptable), different types of disruptions (internal, external), disruption attributes (duration, severity, patterns), adaptive characteristics (capacity, structure, behavior), varying levels of system observability (hidden states), and the need to perform context-sensitive trade-offs (model based, evidence driven). Madni and Sievers [51] offer a formal model-based approach that enables the development and evaluation of mechanisms that protect systems against unpredictable, external disruptions.

Contract-based design (CBD) is a formal method for specifying system requirements, behaviors, and implementations. A contract comprises a pair of invariant assertions, a statement of an input condition, and a guaranteed system behavior under those conditions. The approach is compatible with proofs of correctness, decomposability, trade-space analysis, and online error monitoring. For these reasons, CBD is well suited for describing large-scale systems and for building fault-tolerant mechanisms. Table 9.1 presents a description of contract-based design using mathematical notation.

These are formal, checkable system representations that have been successfully used for defining and validating error detection mechanisms.

Table 9.1 Contract-based design

• A contract, C, is defined by a pair of assertions, C = (A, G), in which A is an assumption made on the environment and G is the guarantee a system makes if the assumption is met.
• For example, a system is guaranteed to produce an output from the set $o \in \{o_o, o_1, \ldots o_{n-1}\} \subseteq O$ when in the state $\sigma \in \{\sigma_o, \sigma_1, \ldots \sigma_{n-1}\} \subseteq \Sigma$ for an input $i \in \{i, i_1, \ldots i_{m-1}\} \subseteq I$ where O is the set of all outputs, Σ is the set of all system states, and I is the set of all inputs [52–54].

Partially observable Markov decision processes (POMDP) are a special case of the Markov decision process (MDP). It is well suited to describing many real-world problem situations that are neither fully observable nor controllable. Interestingly, the Markov assumption is often valid for real-world problems and systems. A POMDP is defined by a set of states, a set of actions, a set of observations, and a transition model, a reward model, and an observation model. The Markov assumption associated with the transition model implies that the optimal policy depends only on the current state. The fact that the operational environment is only partially observable means that the current state is not necessarily known and, therefore, an autonomous agent cannot execute the optimal policy for that state. The general idea is that based on the belief state, the agent can act, judge the impact of that action, and adjust the response accordingly.

Incorporating Flexibility in POMDP Flexibility can be incorporated in POMDP by (a) relaxing the time invariance restriction on the state space and/or action space; (b) adding an evaluation metric to determine best action; (c) updating emission and transition properties of hidden states; and (d) adding the concept of time. By relaxing the time invariance restriction on the state space and action space allows the model to adapt to actual system behavior. Relaxing this restriction also allows state transition probabilities to be adaptable while allowing for emergent states and accounting for unobservable and uncontrollable states. The addition of an evaluation metric allows the determination of the best action when the system is believed to be in a particular state under a certain set of assumptions. Updating emission and transition probabilities determines which outputs are observable and what actions are performed. After sufficient model "training," high confidence can be developed in resulting model probabilities. The model emits observable outputs and internally used action controls. Finally, to add the concept of time, we specify the number of observations before a transition that can occur can be specified, and the evaluation and parameter estimation functions used in model training can be appropriately modified.

Resilience Contract In a traditional contract, an implementation is said to satisfy a design contract if it fulfills guarantees when the assumptions are true. In CBD (Table 9.1), assertions A and G are invariants in which assumptions are preconditions and guarantees are post-conditions. A RC extends invariant contracts in three ways: (a) models unobservable states, (b) looks for emergent behavior, and (c) employs the triggered action itself as a contract. While subsuming fault tolerance, a RC employs flexible assertions that are adaptable and that can also be probabilistic. This feature allows a RC to respond dynamically to unexpected disruptions. A RC adds flexibility to deterministic contracts to generate adaptive system response to a disruption. This characteristic is at the heart of resilience mechanisms. A RC exploits in-use learning and is capable of uncertainty handling and pattern recognition. It is also important to recognize that a traditional contract is not adaptive in the sense that it is defined by a set of invariant assertions and preconditions that fully define the domain of legal inputs and post-conditions. The post-conditions are either correct for legal inputs, or an error is declared. However, for self-adaptive models, we define a new concept that allows for incomplete specification of legal inputs and a flexible definition of post-condition correctness. These characteristics are the essence of a resilience contract (RC).

Fundamental Concepts Classical fault tolerance is represented by "inflexible" assertion. That is, in classical fault tolerance, we make assertions in which assumptions are based on a priori analysis of disruptive events. The contract then guarantees what we believe best corrects the problem situation or leaves the system in a safe configuration. The primary issue, however, is that in a complex system, system state is seldom known. Therefore, we cannot know with certainty whether an action we take will improve or degrade system operation. Consequently, our approach addresses uncertainty by evaluating the impact of small decisions that incrementally move a system from an unknown/problematic state to a healthy or safe state.

Resilience Contract in Operation Let us assume that the complex system appears to be operating reasonably for the most part, but there are indications that not everything is fine. The questions that arise are: Is the system operating okay? Are there signs that the system could be headed for trouble, if certain actions are not taken? Could the system be already in trouble, but we are unaware of that? To answer such questions, we need flexibility in assertions. Rather than have fixed assumptions about faults, typical of fault tolerance, we employ a Partially Observable Markov Decision Process (MDP) to evaluate beliefs about system states. Belief states are initialized with design values and updated during online learning. A policy (which is learned and updated during system operation) based on belief states directs responses which might include: continue to monitor (collect more data), take an action (assert a response and observe what happens), and safe the system and solicit help.

Making the Right Choice If the choice is right, i.e., if the policy directs the correct action, then there will be a discernible or no discernible improvement in system condition. The latter could imply that the system is stable despite some indications to the contrary, and so continued monitoring is appropriate. Also, correctness of choice can be reinforced by increasing confidence in the choice. This is accomplished by changing transition parameter probabilities in the belief model and/or reward parameters for actions.

Making the Wrong Choice If the choice is wrong, then either the belief state is wrong, or the policy is wrong. In either case, a new belief state can be computed, and the action directed by the policy at that state taken. The belief state is updated as needed to reflect new knowledge regarding the system state, and the policy is changed accordingly. In the event that a number of attempted actions do not work, a general safety net response can be pursued.

Verification Because the approach relies on in-use training, it is not possible to verify that the model covers all possible disruptive events. However, rigorous modeling of contract assertions will enable consistency and reachability checks. In hierarchical resilient systems, a key challenge is to assure that resources necessary for specific actions are available when needed and that the different levels within the hierarchy cooperate. Research is needed to fully explore the option space that accomplishes the checking, resource availability, and conflict avoidance goals. In many ways, this problem is similar to that tackled by linear temporal logic (LTL) methods that rely on Büchi automata (BA) for checking. A concept currently under consideration modifies the BA to include hidden states and unknown transition probabilities [4, 50]. A common checking paradigm verifies that the intersection of the language produced by a BA that represents the system, and the language provided by a BA representing the complement of system assertions, is null. Research is needed to evaluate the effectiveness of this approach under the condition of probabilistic and hidden assertions and explore other methods, as necessary. Intelligent control of heterogeneous UAV swarms, a problem of significant interest within both the DOD and the commercial sector, is used to illustrate the feasibility, generalizability, and scalability of the approach.

9.7 Illustrative Example: Resilient UAV Swarm

In this example, the complex system is a swarm of small unmanned aerial vehicles (UAVs) that is tasked to fly in formation to a particular location, collaboratively perform the mission, and return to the command post. The capabilities required to perform the mission are distributed and shared among members of the heterogeneous swarm, rather than colocated within each multi-sensor, multifunction UAV. The overall impact of this architecture is lower cost, smaller UAV size, and *on-demand adaptability* to respond to disruptive events. The latter capability is essential for a resilient swarm, in which each UAV self-adapts to unexpected conditions en route and during the conduct of the mission. In this example, let us assume that the UAV swarm is capable of fine-grain planning and adaptation to accommodate disruptive events. In this system concept, higher-level plans and goals are determined by an external entity (e.g., mission headquarters), while the swarm self-determines local adaptations with the greatest likelihood of achieving higher-level goals in response to perceived threats and other disruptive events. This overall concept is similar to that of driving a rover on Mars. In this concept, the ground station determines high-level goals, while the rover samples the environment and, based on those observations, determines how best to achieve the goals. The primary difference between the Mars rover concept and the resilient swarm is that the swarm is capable of self-reconfiguring resources based on mission goals, when a failure occurs. This resilience capability is achieved through the design of resilience contacts (RCs), which are self-adaptive, *flexible design contracts* that include *probabilistic assertions* and *actions*.

UAV swarms are an area of study both within the DOD and the civilian sector (e.g., agriculture, search and rescue, locating and tracking chemical clouds). They also provide a platform for investigating complex adaptive systems. For example, Scheutz et al. [55] investigated a UAV model for locating and tracking chemical clouds.

Table 9.2 Agent model rules as "contracts"

• If chemical, then activate attraction beacon
• If no chemical, then deactivate attraction beacon
• If UAV within collision range, then turn away
• If no UAV within collision range, turn right (left)
• If no UAV within collision range, update turn decision

Their concept exploits biologically inspired agents that exhibit and accommodate emergent behaviors as the swarm maneuvers toward its goal, while individual UAVs collaborate to avoid collision. The agent model comprises six rules expressed as "contracts" (Table 9.2).

Another relevant example is the development of multi-agent swarm system for distributed automatic target recognition (ATR) using UAVs [56]. This work exploits an ant colony algorithm that uses a "pheromone" trail for directing UAVs to a high-value location (Fig. 9.2). In an ant colony, there is chemical signaling to relay particular conditions to the rest of the group. In keeping with this analogy, we employ messages to inform the rest of the group. Specifically, each UAV (specialist) cooperates with other UAVs (specialists) to accomplish mission objectives. In the traditional ant colony example, the specialist ants either defend or cooperate with other ants that forage or look after the queen. In the UAV swarm example, there are UAVs that look for potential conflicts, or ensure cost-effective arrival at the destination, and/or perform housekeeping functions such as communicate with spacecraft and ground station.

Wei et al. [57] present and discuss a simulation framework for UAV swarm configuration and mission planning. This simulation framework accommodates mission goals, individual positions, and adaptable trajectories. Recently, the US Navy demonstrated an autonomous UAV swarm concept under ONR-sponsored research (http://defensesystems.com/articles/2015/04/15/onr-locust-swarming-autonomous-uavs.aspx). An example of a heterogeneous UAV swarm is shown in Fig. 9.3. The swarm comprises Command and Data Handling (CDH) UAVs and mission UAVs that communicate through a wideband (WB) uplink and narrowband (NB) Tracking, Telemetry, and Command (TTRC). The CDH UAV is capable of communication; real-time plan changes; plan execution and adaptation; swarm formation, control, and status; and coordinated emergency response. The distributed functionality for this UAV type encompasses space and ground communication radios, onboard plan, mission data collection, mission data transmission, swarm communication, swarm control, swarm Health Status and Accountability (HSA), swarm resilience, swarm alarm and status, Attitude Determination and Control System (ADCS), and environment analysis and response. The mission UAVs are capable of imaging and image processing, threat evaluation, threat response, targeting, and damage assessment. Their distributed functionality encompasses ground optical imaging, ground radar, ground IR, sensor fusion and processing, weather sensors, threat sensors, air-to-air defense, air-to-ground defense, CDH drone communication, and ADCS.

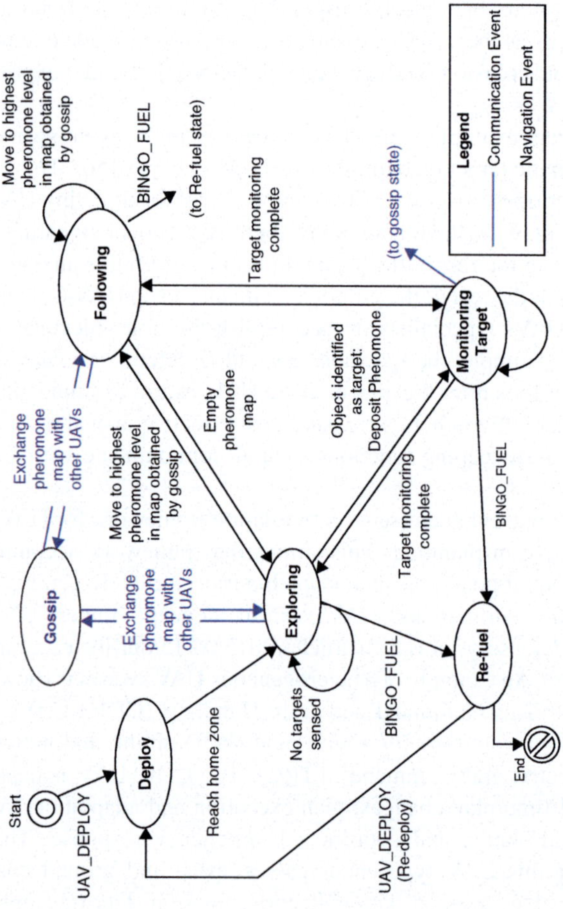

Fig. 9.2 Multi-agent swarming system for distributed ATR

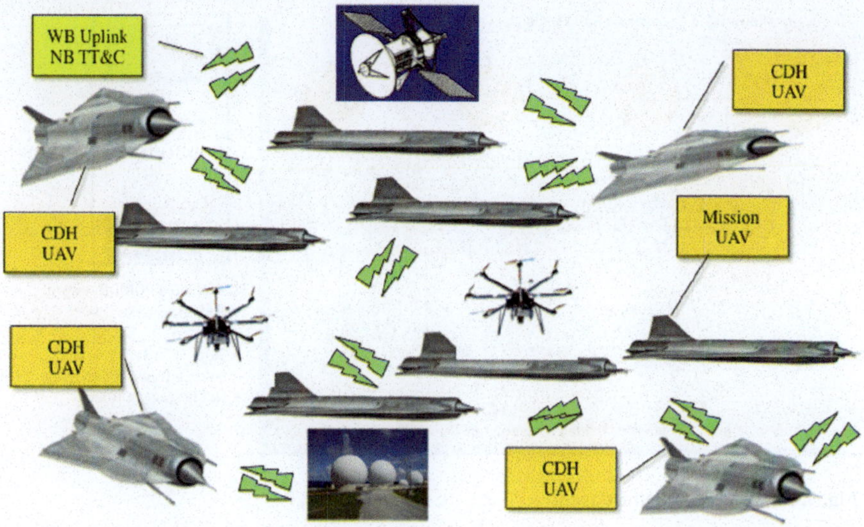

Fig. 9.3 Example of a heterogeneous UAV swarm

The goal of the heterogeneous swarm is to fly to the target using a pre-loaded flight path, and acquire and transmit sensor data for use by command and control (C2). The mission plan is defined by a set of resilience contracts (RCs). RCs enable the swarm to evaluate its current state and determine which options have the best outcome, given current state and likelihood of success. We employ POMDP as the basis for RC. As noted earlier, the POMDP is a decision process in which system dynamics are assumed to follow a Markov decision process (MDP) with some unobservable or hidden states. It is a memoryless decision process based on evaluating transition rewards. The states in a POMDP are unobservable (i.e., hidden) because of uncertainties in knowing the outcome of an action and uncertainties in the knowledge of the environment because of imperfect measurements.

As noted earlier, flexibility is achieved through the introduction of POMDP representation within the RC agent (Fig. 9.4). As shown in this figure, a RC agent has deterministic and probabilistic parts. The agent accepts inputs and generates observable outputs while operating within a partially observable environment. The RC agent evaluates the POMDP reward and responds accordingly. Typical responses are to keep going, stop, or enforce trajectory to a safe state and notify support team.

The first step in modeling is creating a POMDP model. An exemplar POMDP is shown in Fig. 9.5. The POMDP maintains a probability distribution over all state transitions and observations. The probability distribution is used to evaluate what is the most likely state of the system and the reward(s) for making such decisions. Typically, an "agent" performs this evaluation and makes decisions based on a "policy." The policy determines which actions to take in a given belief state. The agent in this case is the RC, while the policies are the basis for decisions.

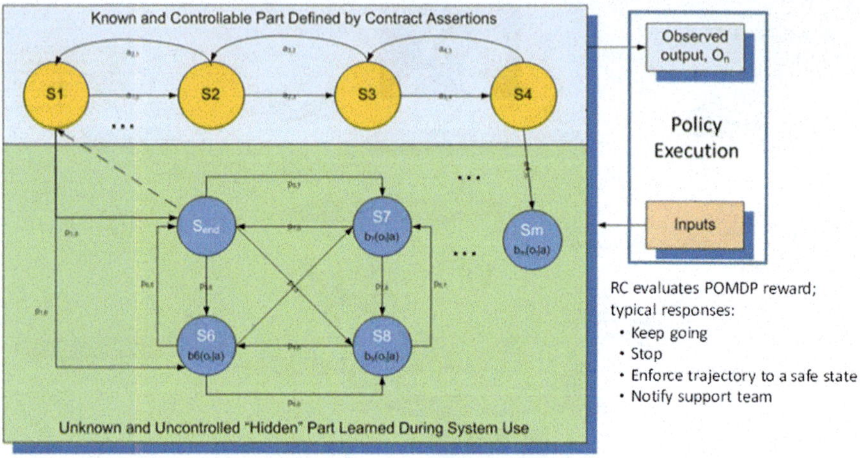

Fig. 9.4 Resilience contract agent

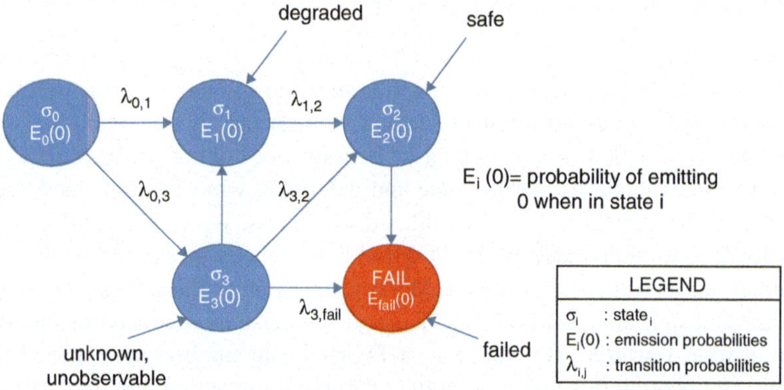

Fig. 9.5 Exemplar POMDP

This model, which includes observable and unobservable states, is trained during system use by determining transition and emission probabilities. During use, system outputs are used to estimate the state that maximizes the posterior probability from the observations as well as the maximum likelihood that the POMDP parameters could have produced the observations. For states known to be problematic or potentially problematic, the POMDP evaluates a reward function that determines whether an immediate action or delayed action is expected to produce the best outcome. Actions are intended to move the system away from problematic states to desirable or safe states. For example, transition $\lambda_{3,1}$ is an example of a transition from a "bad" state to a good state. The next question is how to accommodate emergent behaviors, i.e., those that were not previously observed. Previously unobserved

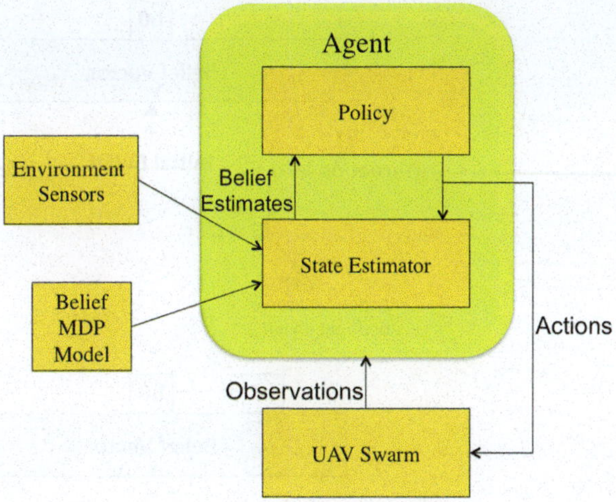

Fig. 9.6 Example swarm control architecture

disruptions or partially observable disruptions can lead to previously unknown states. These states can give rise to emergent behavior. Emergent behavior, due to previously unknown states, is evaluated by computing the statistical distance to prior known states. When the weighted distance (i.e., Mahalanobis distance) exceeds a limit, a new state is added. This algorithm [58] avoids conditions in which the POMDP predicts a strong match to a prior known state even though the observation mean is far from that state. Figure 9.6 presents an example of the swarm control architecture.

As shown in this figure, swarm control is based on creating an optimal policy based on belief estimates provided by the state estimator. The state estimator relies on observations from the UAV swarm, environment sensors, and MDP belief model to generate updated belief estimates. Policy actions act on the UAV swarm and are used by the state estimator to update state information.

A simple CONOPS for a UAV swarm is used to illustrate the approach. In this simple scenario, the UAV swarm needs to turn left or right to avoid an obstacle. There is uncertainty regarding the location of the threat. The threat could be to the left or the right of the swarm. A decision needs to be made to veer left or veer right. If the swarm veers right and the threat is located/headed to the right, dire consequences can result. The same is true if the swarm veers left, and the threat is located or headed to the left. There are three possible actions that the swarm can take: veer left; veer right; continue flying straight ahead while continuing to collect more data on the threat. The POMDP policy for this simple CONOPS has to deal with a variety of considerations such as UAVs do not inadvertently crash into each other; all UAVs get safely to their destination; UAVs avoid potentially disruptive events; and if one (or more) UAVs is shot down, the remaining UAVs need to reorganize and reallocate functionality to ensure achievement of objective to the extent feasible. The key ideas behind an optimal POMDP policy are twofold: A POMDP policy maps current belief into an action, and an optimal POMDP policy is a continuous

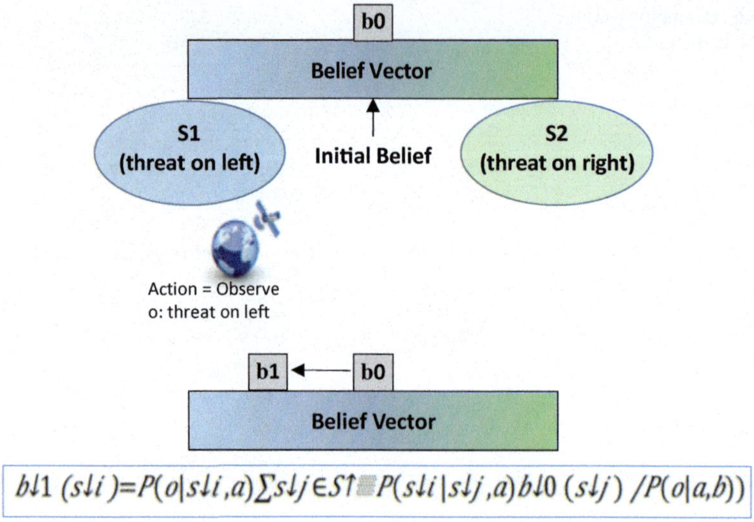

Fig. 9.7 Iterative update of beliefs

solution of a belief MDP. Figure 9.7 shows the equation for summation of outcomes based on the path the UAVs take. The equation normalizes the rewards and penalties. As shown in Fig. 9.7, the system starts with a 50–50 belief that the threat could be to the left or the right. The system then makes an observation of a potential threat to the left. The system revises its belief from b0 to b1, i.e., there is a greater belief that the threat could be to the left. The belief is updated in accord with recursive Bayesian Belief Network using new observation and current state.

A key problem with such state space models is that they are subject to combinatorial explosion. To contain this explosion, several methods can be potentially applied including pruning (Bellman equation), branch and bound, heuristic search, Monte Carlo search, and policy tree.

9.8 Success Metrics and Desired Outcomes

The metrics that characterize the engineering of resilient systems span both the product and process perspectives. The key metrics pertain to the scope and types of design adaptability, the speed and efficiency of engineering iterations, and the ability to inform decisions of mission needs in timely fashion. Table 9.3 presents the detailed measures associated with these metrics.

The envisioned end state of the overall methodology has three desired outcomes: improved (more resilient) system design; improved engineering processes; and improved engineering and design capabilities (i.e., tools). Figure 9.8 presents the envisioned end state associated with the proposed methodology.

Table 9.3 Success metrics

• Design Adaptability (robustness metric) – accessibility and modifiability of diverse system models – reuse and replacement potential (modularity) – interoperability potential (standardized interfaces, compatible semantics) – continuous analyzability (performance, vulnerabilities, trust)
• Engineering Iteration Scope, Rate, and Efficiency – ability to integrate 3D geometry, electronics, software in virtual design – how early problem is discovered – how short cycle times are for risk reduction phases with prototypes – how fast cycle times are for design / build / test
• Information Tmelines for Decision Making – depth and number of options generated – breadth of trade space analyses (CONOPS, environment) – depth of collaboration scope (interaction, iteration) – ability to simulate and experiment in synthetic operational environments

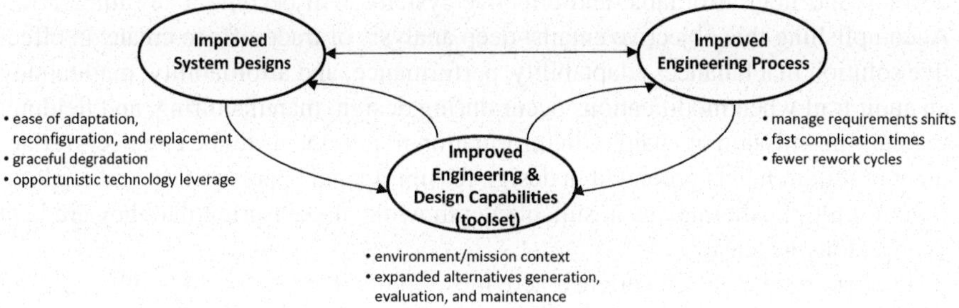

Fig. 9.8 Envisioned end state: three desired outcomes

As shown in Fig. 9.8, the expected overall outcome is a technology-enabled methodology for engineering of resilient systems. This methodology can be implemented and evaluated through a series of pilots geared to each challenge area.

9.9 Summary

Engineered resilience, an essential characteristic of systems in the twenty-first century, is a science and technology problem, rather than a process compliance or reengineering problem. The hard problem from a customer's perspective is defining what customers need, not what they say they want. Trade-space analysis is an essential capability for engineering resilient systems. Instead of pursuing an optimal design, trade-space analysis focuses on the proper formulation and balancing of system needs and stakeholder requirements. When needs are prematurely translated into requirements or key performance parameters without expanding and exploring the option space, both the engineering process and the end product are adversely impacted. By basing cost and schedule on unwarranted assumptions of technology feasibility, we risk schedule delays and cost overruns. We also limit the full consideration of technology-enabled opportunities.

Resilient systems require a well-integrated development process, in which the effectiveness of the process is determined by the frequency and quality of interaction (i.e., communication) among the different aspects of a process. Today, design begins long after requirements have been specified. In reality, the two processes need to be performed near-simultaneously in tight coordination with each other to realize effective systems.

In the future, rapid and inexpensive upfront engineering will be needed to secure program managers' buy-in for decisions made with engineering rigor. Upfront engineering will need to: automatically consider multiple variations; propagate changes and maintain constraints; introduce and evaluate multiple usage scenarios; explore technology and operational trade-offs; iteratively refine requirements; adapt and build in adaptability; and learn and update. The creation of new ways to rapidly develop and field affordable and effective systems will become a key imperative. Accomplishing this objective entails: deep analysis of trade-offs to create an effective solution that balances adaptability, performance, and affordability; maintaining an audit trail when modifications occur during design, manufacturing, and fielding; regular transmission of changes in information requirements to the engineer; refinement of requirements grounded in design feasibility and opportunities; and exploration of multiple alternatives in suitable depth while also ensuring that they are kept open as long as feasible.

Performing the aforementioned activities rapidly, repeatedly, and with sufficient flexibility to adapt to new technologies will be essential. Examples of these new technologies are semantically expressive models to analyze many more designs than was previously possible; exploiting knowledge of operational context in system design and test; and uncertainty handling and risk analysis tools to manage combinatorics of in-depth analyses of trade-offs and options during systems engineering. And, finally, Model-Based Engineering (MBE) and Platform-Based Engineering (PBE) approaches, when combined with substantial computing power, are important enablers of rapid trade-space exploration, as well as reduction in development, cycle times, and life-cycle costs – the key enablers of resilient systems.

References

1. Neches R, Madni AM. Towards affordably adaptable and effective systems. Syst Eng. 2012;16(2):224–34.
2. Goerger SR, Madni AM, Eslinger OJ. Engineered resilient systems: a DoD perspective. Procedia Computer Science. 2014;28:865–72.
3. Madni AM, Sievers M. A flexible contract based design framework for evaluating system resilience approaches and mechanisms IIE annual conferenec and expo May 30–June 2 Nashville, 2015.

4. Sievers M, Madni AM. A flexible contracts approach to system resilience. In: Systems, man and cybernetics (SMC), 2014 IEEE international conference on. San Diego: IEEE; 2014. p. 1002–7.
5. Edwards T. Architecting for resiliency: army's common operating environment (COE). DTIC document, 2011.
6. Madni A. Integrating humans with and within software and systems: challenges and opportunities. (Invited Paper) Crosstalk, J Def Softw Eng. 2011;5:4–8.
7. Neches R. Engineered resilient systems (ers) s&t priority description and roadmap. 2011.
8. Welby S. DoD systems engineering update. National Defense Industrial Association (NDIA) Gulf Coast; 2010. p. 1–25.
9. Carter AB. Should-cost and affordability, memorandum for defense acquisition and logistics professionals. Washington, DC, Office of the Under Secretary of Defense (Acquisition, Technology, and Logistics), 2011.
10. Kendall F. Better buying power 2.0: continuing the pursuit for greater efficiency and productivity in defense spending. Memorandum to the Defense Acquisition Workforce Department of Defense, Office of the Undersecretary of Defense, Acquisition, Technology, and Logistics, Washington, DC, 2012;13.
11. De Neufville R, Scholtes S. Flexibility in engineering design. Cambridge: MIT Press; 2011.
12. De Neufville R, Scholtes S, Wang T. Real options by spreadsheet: parking garage case example. J Infrastruct Syst. 2006;12(2):107–11.
13. Copeland TE, Antikarov V. Real options: a practitioner's guide. New York: Texere; 2001.
14. Page J. Flexibility in early stage design of US navy ships: an analysis of options. J Ship Prod Des. 2012;28(3):128–33.
15. Wu Y-T. Computational methods for efficient structural reliability and reliability sensitivity analysis. AIAA J. 1994;32(8):1717–23.
16. Hammer M, Champy J. Reengineering the corporation: manifesto for business revolution, a: Zondervan; 2009.
17. Hammer M. Beyond reengineering: how the process-centered organization is changing our work and our lives 1900.
18. Davenport TH. Process innovation: reengineering work through information technology. Boston: Harvard Business Press; 2013.
19. Davenport T. Re-engineering - the fad that forgot people. New York: Fast Company; 1995.
20. Drezner JA, Jarvaise JM, Hess RW, Hough PG, Norton D. An analysis of weapon system cost growth. DTIC Document, 1993.
21. Christensen DS, Stearle DA, Vickery C. The impact of the Packard Commission's recommendations on reducing cost overruns on defense acquisition contracts. DTIC Document, 1999.
22. Madni AM. A process-centered methodology for engineering of complex systems. Fredricksburg: Dahlgren Division Naval Surface Warfare Center; 1994. p. 245.
23. Madni AM, Brenner M, Costea I, MacGregor D, Meshkinpour F, editors. Option generation: problems, principles, and computer-based aiding. Proceedings of the 1985 international conference on systems, man, and cybernetics, Tuscon, 1985.
24. Madni AM. Generating novel options during systems architecting: psychological principles, systems thinking, and computer-based aiding. Syst Eng. 2013;17(1):1–9.
25. Madni AM. Elegant systems design: creative fusion of simplicity and power. Syst Eng. 2012;15(3):347–54.
26. Madni AM. Adaptable platform-based engineering: key enablers and outlook for the future. Syst Eng. 2012;15(1):95–107.
27. Zarboutis N, Wright P, editors. Using complexity theories to reveal emerged patterns that erode the resilience of complex systems. Proceedings of the second symposium on resilience engineering, Juan-les-Pins, 2006.
28. Di Marzo Serugendo G, Fitzgerald J, Romanovsky A, Guelfi N, editors. A metadata-based architectural model for dynamically resilient systems. Proceedings of the 2007 ACM symposium on applied computing; ACM, Seoul, Korea; 2007;566–572.

29. Ward P, Suss J, Eccles DW, Williams AM, Harris KR. Skill-based differences in option generation in a complex task: a verbal protocol analysis. Cogn Process. 2011;12(3):289–300.
30. Woods DD. Essential characteristics of resilience. In:Resilience engineering: concepts and precepts. Aldershot: Ashgate Publishing; 2006. p. 21–34.
31. Kichkaylo T, Roesler G. Automating trade space analysis in systems design. Conference on systems engineering research, Los Angeles. 2011.
32. Madni A, Moini A, Madni C. Cross-cultural decision making training using behavioral game-theoretic framework. Advances in cross-cultural decision making: Informa UK Limited; 2010. p. 385–94.
33. Madni AM. Integrating humans with software and systems: technical challenges and a research agenda. Syst Eng. 2010;13(3):232–45.
34. Deevy E. Creating the resilient organization: a rapid response management program. Englewood Cliffs: Prentice Hall; 1995.
35. Levin C. Levin, McCain introduce legislation to reform weapons systems acquisition process. Michigan, Carl Levin, United States Senator [Internet]. 2009 2010. Available from: http://levin.senate.gov/newsroom/release.cfm?id=308529.
36. Pal N, Pantaleo D. The agile enterprise: reinventing your organization for success in an on-demand world. New York: Springer Science & Business Media; 2005.
37. Sheffi Y. The resilient enterprise: overcoming vulnerability for competitive advantage, vol. 1. Cambridge: MIT Press Books; 2005.
38. Hind P, Frost M, Rowley S. The resilience audit and the psychological contract. J Manag Psychol. 1996;11(7):18–29.
39. Holling CS, Gunderson LH. Resilience and adaptive cycles. In:Panarchy: understanding transformations in human and natural systems. Washington, DC: Island Press; 2016. p. 25–62.
40. Leveson N, Dulac N, Zipkin D, Cutcher-Gershenfeld J, Carroll J, Barrett B. Engineering resilience into safety-critical systems. In:Resilience engineering–concepts and precepts. Aldershot: Ashgate; 2006. p. 95–123.
41. Jackson S. Architecting resilient systems: accident avoidance and survival and recovery from disruptions. New York: John Wiley & Sons; 2009.
42. Sullivan PE. The future for naval engineering: Topics for the research and development community, paper prepared for the Committee on naval engineering in the 21st century, transportation research board, national academy of science. Washington, DC. 2011.
43. Westrum R. A typology of resilience situations. In:Resilience engineering: concepts and precepts. Aldershot: Ashgate; 2006. p. 55–65.
44. Madni AM, Jackson S. Towards a conceptual framework for resilience engineering. IEEE Syst J. 2009;3(2):181–91.
45. Fiksel J. Designing resilient, sustainable systems. Environ Sci Technol. 2003;37(23):5330–9.
46. Holling CS. Engineering resilience versus ecological resilience. In:Engineering within ecological constraints. Washington DC: National Academy Press; 1996. p. 31–44.
47. Mallak LA, editor. Toward a theory of organizational resilience. Management of engineering and technology, 1999 technology and innovation management PICMET'99 Portland international conference on July 25–29, 1999, IEEE, Portland. 1999.
48. Amin SM, Horowitz BM. Toward agile and resilient large-scale systems: adaptive robust national/international infrastructures. Glob J Flex Syst Manag. 2008;9(1):27.
49. Gunderson LH, Pritchard L. Resilience and the behavior of large-scale systems. Washington DC: Island Press; 2012.
50. Madni AM, Sievers M, editors. A flexible contract-based design framework for exaluating system resilience approaches and mechanisms IIE annual conference and expo 2015 May 30–June 2 Nashville, 2015.
51. Madni AM, Sievers M. Model based systems engineering: motivation, current status and needed advances systems engineering, SAE MBSE Report. 2015
52. Sangiovanni-Vincentelli A, Damm W, Passerone R. Taming Dr. Frankenstein: contract-based design for cyber-physical systems. Eur J Control. 2012;18(3):217–38.

53. Meyer MA, Booker JM. Eliciting and analyzing expert judgment: a practical guide. Philadelphia: Siam; 2001.
54. Le Traon Y, Baudry B, Jezequel JM. Design by contract to improve software vigilance. IEEE Trans Softw Eng. 2006;32(8):571–86.
55. Scheutz M, Schermerhorn P, Bauer P, editors. The utility of heterogeneous swarms of simple UAVs with limited sensory capacity in detection and tracking tasks. Proceedings 2005 IEEE swarm intelligence symposium, 2005 SIS 2005; IEEE, Pasadena, CA; 2005.
56. Dasgupta P. A multiagent swarming system for distributed automatic target recognition using unmanned aerial vehicles. IEEE Trans Syst Man Cybern Syst Hum. 2008;38(3):549–63.
57. Wei Y, Blake MB, Madey GR. An operation-time simulation framework for UAV swarm configuration and mission planning. Procedia Comput Sci. 2013;18:1949–58.
58. Lee S, Li L, Ni J. Online degradation assessment and adaptive fault detection using modified hidden markov model. J Manuf Sci Eng. 2010;132(2):021010.

Chapter 10
Autonomous System-of-Systems

> *"There is nothing more powerful than an idea whose time has come."*
>
> –Victor Hugo, poet

10.1 Road to the Present

"Autonomous" means the ability for self-governance and independent operation. Autonomous vehicles (AVs) are required to exhibit requisite performance for extended durations with the desired level of reliability and safety in the face of significant environment uncertainties, malfunctions, and disruptions without external intervention. AVs today are network-enabled. Soon they will be capable of communicating with other vehicles and smart structures in the immediate vicinity (for collision avoidance), and distant vehicles and structures (for congestion management). Most AV concepts employ a network connection to the cloud, other vehicles, and buildings. Most AV concept of operations today do not make autonomous decisions with respect to destination and route selection.

Consumer acceptance of AVs will depend to a large extent on how they are introduced in the real-world environment. For example, AVs without a steering wheel, and eventually without a human-in-the-loop, are radical concepts that need to be acceptable to passengers. Fortunately, there is a precedent. Elevators, when first introduced in office buildings, were operated by elevator operators who were responsible for making sure that elevator cars stopped at specific passenger-requested floors. Elevator operators would open and shut elevator car doors at different floors and, in the process, made occasional mistakes. However, such infrequent mistakes did not deter passengers in a hurry to leap into elevator cars moments before the elevator operator shut the door.

It did not take long for the elevator industry to respond. They added doors with safety bumpers and automated stopping. In 1900, elevator cars became "self-driving." Passengers felt uneasy initially, with some routinely stepping out of the elevator car expecting to locate an elevator operator! It did not take long for the elevator industry to conclude that people needed an entirely *different experience* if they were to trust elevators and start viewing them as safe. The elevator industry started an advertising campaign showing children and grandmothers confidently

A.M. Madni, *Transdisciplinary Systems Engineering*,
DOI 10.1007/978-3-319-62184-5_10

boarding elevators, pushing buttons, and riding elevator cars. They even piped in reassuring instructions to passengers to press the right button on the panel to get to desired floors. The instructions also drew their attention to the emergency "stop" button to explain its use. These simple enhancements worked and produced valuable lessons.

Today the AV industry is exploiting these lessons. An AV today is capable of driving to its destination without human interaction and with and without human occupants. Equipped with a sensor suite, preprogrammed logic, connection to the Internet of Things (IoT), and deep machine learning algorithms, AVs are rapidly becoming capable of assessing and reacting to various traffic situations involving other moving vehicles (including bicyclists and motorcyclists) and pedestrians. Today AV designers such as Google are specifically focused on shortening the time between AV introduction and consumer acceptance.

10.2 Trends in Autonomous Vehicles

At the time of writing this book, the development of connected AVs has made significant progress. Players in this market segment include *traditional automotive companies* (e.g., General Motors, Ford, Mercedes, BMW, Audi, Toyota, Volvo), *technology companies* (e.g., Uber, Google, Apple), and *component technology providers and startups*. Some of the more important technologies being acquired for connected AVs are infrastructure and connectivity, cybersecurity in connected vehicles, computer vision and 3D sensing, mapping and machine learning, sensor and systems integration, software updates using over-the-air technologies, path and motion planning, video and image processing, end-to-end mapping, emergency vehicle detection on roadways, and automated driving and decision making.

Startups are being acquired at a torrid pace by global players in the AV market. Israel has become a breeding ground for startups in this space. Without a local market for self-driving cars, Israel is focused on providing component technologies to global players in this space.

Today a few key trends are driving AV competition: radically new, affordable technologies; new component and niche technology entrants; new mobility concepts; and new regulations and policies [1].

Radically New, Affordable Technologies These include fifth-generation wireless technology that is able to stream data from the cloud in near real-time, faster processing to execute AI algorithms and steer self-driving vehicles, complex and low-cost sensors that make vehicles context-aware, and integration software.

New Component and Niche Technology Entrants Today nontraditional technology companies are beginning to disrupt the technology value chain by offering innovative products. For example, NVIDIA makes systems for dashboard functions and autonomous driving and mapping. Other new entrants are committed to testing new ideas and accelerating product development. Their data-centric business models are a significant departure from the business models of traditional automakers. These new models are far more dependent on revenues from ongoing services and sale of information. These new entrants can potentially impact not just vehicle operation, but an entire industry.

New Mobility Concepts There is a definite movement toward ride-sharing driven in large part by dramatic reductions in transportation costs associated with connected cars.

New Regulations and Policies While policies and regulations typically lag technology advances, regulators can be expected to issue new laws and regulations to ensure the safety of self-driving cars and its passengers, once the right combination of technologies gets adopted.

These trends partly explain the investments being made by automobile manufacturers in connected technologies. They also explain why data-centric technology companies such as Google, Uber, and Apple want to succeed in the self-driving vehicle space. To succeed in this ultracompetitive, dynamic space requires a change in mindset. To begin with, connected, autonomous cars need to be viewed as a *platform*, comprising a set of technologies that will inevitably change the core business model. In other words, AVs should not be viewed as a product, but as a platform. Furthermore, the focus on the typical 7-year development cycle for a new car is seeing a shift to cloud-based services that are developed in mere months and deployed in near real-time to customers worldwide [1].

Cybersecurity continues to be a daunting and ongoing challenge for connected cars, with hackers continuing to break into automobiles to take over their navigation, control, and safety functions. Break-ins in a connected car universe can disrupt not only entire traffic flows but compromise an entire fleet of cars and illegally acquiring personal data flowing between a vehicle and the cloud through vehicle-based consumer apps and services. Determined hackers are capable of working their way into the IT system of the car's OEM, suppliers, and service providers. It these scenarios appear dire, it is because they are. It should not come as a surprise that connected vehicles are going to be vulnerable to security breaches. After all, they are complex systems comprising multiple, interconnected digital systems, with potential vulnerabilities in their interactions. Today security issues can also be traced to the difficulty in orchestrating the work of multiple participants engaged in producing networked AVs.

However, in this regard, certain preemptive measures can be taken to secure connected cars and dramatically reduce security breaches. Some of the more effective measures include:

- Embedding security in the development process
- Employing remote software updates through connectivity functions
- Incorporating risk management, progress monitoring and reporting, and incident management throughout the software development life cycle

However, the security problem cannot be handled by merely incorporating security measures into connected cars [1]. The software also needs to be thoroughly tested. Testing a complex software-intensive system is no easy task. However, progress is being made in the use of lightweight formal modeling methods to help with both verification and testing. However, the problem does not end here because new threats will inevitably appear over time, necessitating periodic update of the security software in coordination with suppliers.

There is also a *psychological* aspect to cybersecurity. While ensuring the safety of a network-enabled AV and its services from hackers is an absolute imperative, cybersecurity has a salutary psychological effect on consumers. It instills confidence and trust in them to return to the dealership to continue to buy self-driving, networked vehicles. Ultimately, securing networked AVs is a collective task that requires multiple stakeholders (e.g., OEM, suppliers, regulators, dealerships, insurance companies) to cooperate. The closer we get to real-world deployment, the more important it becomes to address these concerns in ways that ensure that cybersecurity solutions will endure.

The need for AI techniques and machine learning is well recognized in the networked, self-driving vehicle community. By building the capability to "learn" in self-driving vehicles, these vehicles will continue to improve their ability to adapt on-the-fly to varying circumstances and react appropriately to new demands and new tasks. While these capabilities are becoming commonplace in robotics and industrial machines with access to sensor and network data, networked AVs are expected to operate in far more complex situations and challenging environments (e.g., traffic congestion, adverse weather, changing passenger cognitive state). So, they will need to be even smarter [1].

Machine learning is becoming an integral part of AVs. Inputs into machine learning systems comprise individual and fused data from onboard sensors, cameras, and laser systems that monitor road situations and traffic conditions. This technology is slated for introduction *within* the vehicle to monitor and analyze the state (i.e., health, awareness) and comfort level of occupants and respond accordingly. In the not too distant future, AI systems will begin to exploit data from external services such as social media, entertainment channels, and smart homes to offer suggestions and items of interest to occupants. Eventually, a single AV will routinely cooperate with other AVs within a system-of-systems (SoS) network to explore new solutions that leverage cloud-based analysis of data and trends.

Advances in AV *human-machine interfaces* include employing data from navigation systems, road condition sensors, and Internet of Things (IoT) to construct situation displays for passengers. Specifically, advances in human-machine interface technology coupled with AI will provide passengers with the status of what the vehicle is doing at any point in time (e.g., attempting to park), while also querying passengers about their intent (e.g., when they wish to be picked up).

AV manufacturers today are striving to make the vision of connected AVs into a reality. They are continuing to experiment with and test various promising technologies, even as they continue to connect the AVs to the world around them. At the same time, they are developing increasingly more sophisticated AV prototypes.

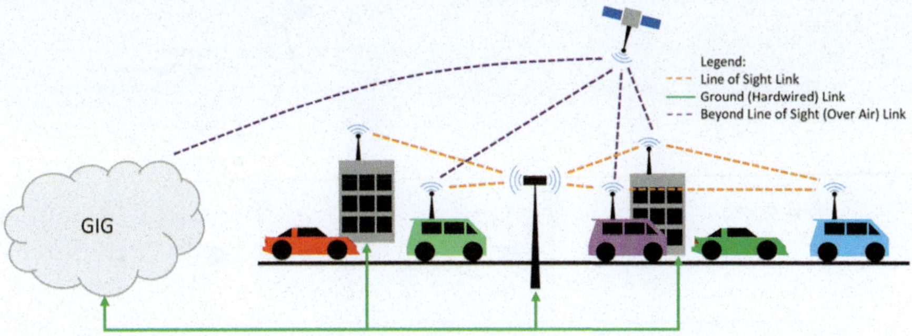

Fig. 10.1 AVs linked through a smart infrastructure form a system-of-systems (SoS) network

Their business models are beginning to change as the market continues to be segmented, with players in each segment striving to provide unique capabilities that complement the technologies of AV manufacturers.

And finally, there are different mindsets at work in this exciting space. While some are focused on bringing the internet into the vehicle and improving vehicle quality, others are focused on offering mobility on demand. The latter see the AV as an integral part of the IoT.

10.3 Connected AVs: A System-of-Systems (SoS) Perspective

Figure 10.1 shows multiple AVs interconnected via a smart infrastructure, that is enabled by the IoT [2]. Each AV is an independent system that leverages the connectivity provided by the smart infrastructure to communicate and share data with other vehicles, structures, and the Global Information Grid (GIG). Therefore, connected AVs can be viewed as a system-of-systems (SoS) network. These definitions inform the definition of operational scenarios and use cases which relate to AV-SoS objectives, stakeholder objectives, and AV-SoS requirements [3]. Collectively, these factors inform SoS Integration and SoS Verification and Validation (V&V).

Note the presence of two legacy vehicles in the environment which are not part of the AV network but are members of the larger SoS.

By 2020, AVs are expected to become an integral part of a SoS network comprising AVs and human-driven vehicles. The transition from human-driven vehicles to AVs can be expected to occur in stages. Prior to all vehicles becoming fully autonomous and networked, there will be a transition period in which networked AVs will coexist with human-driven stand-alone vehicles. This stage in the evolution of networked AVs is also depicted in Fig. 10.1, which shows stand-alone human-driven vehicles being operated alongside networked AVs.

In Fig. 10.2, vehicles (agents) form line-of-sight (LOS) wireless connections via various techniques (e.g., Wi-Fi, cellular, radio). These connections form local, ad hoc meshes, with infrastructure elements having hardwired connectivity to the GIG

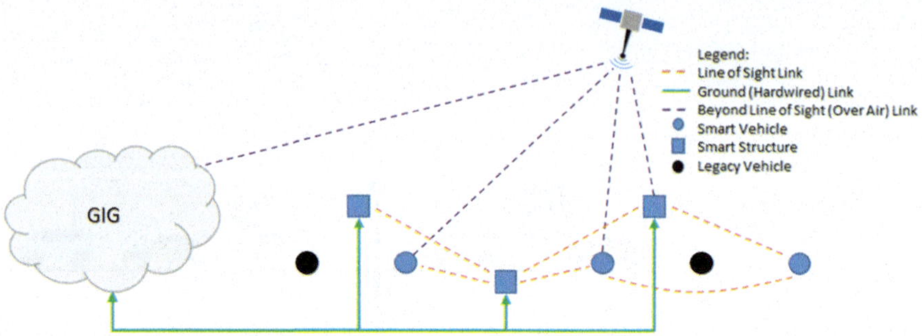

Fig. 10.2 Nodal depiction of AV-SoS in Fig. 10.1

(e.g., cable/DSL, fiber optics). Some elements have beyond line-of-sight (BLOS) connectivity via specific hardware (e.g., satellite communication links or Iridium networks). The BLOS link also connects to the GIG.

Not depicted in this figure are drivers of human-driven vehicles who may carry mobile phones or other devices that wirelessly connect to cell towers, Wi-Fi devices, or other smart devices in the environment. These connections can effectively transform "dumb" entities into "smart" entities (e.g., Think about how smartphones pull emergency broadcast alerts). The challenge is either mandating or getting people to opt into pushing their data to the local mesh instead of just pulling alerts from it. If they do not push the required data, then that data has to be sensed and modeled by neighboring smart agents to support predictive analytics. Since at this point we do not know what, if any, capability will exist in this regard, this information is intentionally left out of this figure. Also not depicted in the figure are airborne agents (e.g., news helicopters) that could relay and extend the local mesh or even connect the BLOS to the GIG.

Increased network flexibility and adaptability enhances link persistence, while also improving situational awareness and learning capabilities. It also illuminates required trade-offs, as the need for additional hardware and software is identified, impacting cost, packaging, and vehicle styling, e.g., most people will not want an antenna farm on their roof! Also, packaging may have to be region specific. For an AV in Los Angeles, Wi-Fi and cellular may suffice. However, an AV in Utah may well require BLOS connectivity.

It is worth noting that the advent of new technologies typically implies new types of failures and accidents [5]. According to Virilio [5], "When you invent the ship, you also invent the shipwreck; when you invent the plane, you also invent the plane crash; and when you invent electricity, you invent electrocution." Such is the case with AVs as well – new capabilities will come with new types of problems and unanticipated accidents. Chatfield characterizes such accidents as "the spots where anticipation met reality and came off worse" [6]. Nevertheless, the opportunity exists to "make tomorrow's worst better than today's worst" [6]. While, on very rare occasions, aircraft may crash and result in fatalities, the use of checklists, fault-tolerance algorithms, and other resilience measures invariably circumvent and prevent the majority of crashes. Furthermore, smart sensors, preprogrammed logic, deep machine learning, and vehicle maneuverability can be expected to prevent the majority of potential accidents.

Despite unprecedented safety and reliability of vehicles used for transport, Chatfield cautions that "more than 1 million people are killed in car accidents around the world each year, and more than 50 million are injured" [6]. These accidents, invariably the result of lapses in attention or poor judgment, show up as human error. This data, in large part, provides the impetus for AVs and explains their appeal today. The viability of AVs as the principal mode of transportation will continue to increase with advances made by automobile manufacturers and hi-tech companies pursuing AVs as a lucrative market. For example, between 2009 and January 2016, Google's test vehicles crashed only 17 times while traversing 1.3 million miles. Each of the 17 crashes was attributed to human error [7]. With AVs, accidents resulting from human error or arising from inattention or distracted driving will be entirely eliminated.

Finally, there is the issue of decision making and ethics in AVs. Ethical reasoning continues to garner increasing attention as AVs move closer to becoming a reality [8–11]. Fortunately, most ethical issues that arise during routine driving are readily resolvable. The extremely rare case when an AV can face an imminent accident, with no good options, is best addressed through government regulations and legislation and avoided in some cases through appropriate infrastructure design. For example, dedicating a lane to AVs the way we do for high-occupancy vehicles (HOVs) is an obvious example of the latter.

10.4 High-Reliability Imperative

Ultimately, AVs need high reliability for consumer acceptance. A *high-reliability system* is capable of relatively error-free operation over extended time periods making consistently good decisions that result in highly reliable and safe operation. Specifically, a high-reliability system solution needs to resist oversimplification while incorporating a context-aware, smart infrastructure that (1) observes and tracks minor disruptions, failures, and anomalies; (2) is sensitive to operations; (3) exhibits resilience in the face of disruptions; and (4) exploits embedded expertise to make informed decisions. The first capability requires "anticipation," while the remaining three require "containment."

Engineers in the aviation, nuclear power, and medical devices industries have in-depth experience with high-reliability systems. The AV industry being relatively new has several open questions with far-reaching implications. These include: a) Will reliability requirements for self-driving cars be similar to those for other high-reliability systems, or should they draw on experiences with human-operated cars that are subject to occasional error-proneness on the part of the driver? b) Would these two very different approaches produce similar or different requirements? If different, which of the two approaches is the more credible and why? Answers to

Table 10.1 Approaches for high-reliability systems

• Analysis using physics-based model – e.g., the lifetime of a solid state junction is based on solid state diffusion of dopants in a semiconductor
• Analysis using end-to-end models of large systems – e.g., derive system failure rate from component failure rates
• Setting safety margins – adopting more stringent requirements for key performance parameters (KPPs) to compensate for unknowns – e.g., a system that is required to last 10 years may be designed to last 20 years
• Redundancy to achieve fault tolerance and resilience – functional redundancy achieving redundancy using dissimilar methods (hardware or software) – physical redundancy achieving redundancy using identical hardware
• Testing – over the range of expected operating conditions (robustness) – beyond the range of expected operating conditions (resilience)
• Manufacturing process control – ensure what is built is the same as what was tested
• Operational limits – preclude system being operated outside the envelope used to test it

such questions need to be dealt with before AVs are introduced in large numbers in operational environments. Table 10.1 presents currently available approaches for high-reliability systems.

An AV consists of mechanical and electrical subsystems, sensor subsystems, and software subsystems that include deep learning subsystems. Each AV subsystem stands to benefit from the approaches presented in Table 10.1.

Mechanical and Electrical Subsystems All approaches presented in Table 10.1 are applicable. These include physics-based modeling, end-to-end model analysis to derive system failure rates, failure margins, redundancy, testing, manufacturing process control, and operational limits.

Sensor Subsystem The methods presented in Table 10.1 are useful to assure the reliability of physical sensors. Sensor subsystems typically include software that interprets the output of physical sensors. The expected rates of false negative and false positives for each sensor subsystem need to be evaluated. This evaluation requires testing all the conditions that can potentially be encountered in actual operation.

Software Subsystems (e.g., High-level Supervisory Control Systems) To the extent that these subsystems are designed to handle a set of "use cases," there is uncertainty about how the subsystem will behave when confronted by a situation that is significantly different from one of the "use cases" used in its design.

Deep Learning Subsystem Since the behavior of the system depends on its cumulative experience, many approaches to reliable system design can be expected to change for these systems.

Table 10.2 An AV network is a SoS

• Operational independence of AVs – AVs operate independently as part of a traffic ecosystem
• Managerial independence of AVs – AVs governed independently while being part of traffic ecosystem
• Evolutionary development of SoS – development and existence are evolutionary with functions and purposes added, removed, and modified with experience and need
• Emergent SoS behavior – AV-SoS performs functions and carries out purposes that do not reside in any single AV – AV-SoS behaviors are emergent – cannot be realized by a single AV
• Geographic distribution – AVs primarily exchange information – not mass or energy

The testing of high-reliability AVs requires answers to the following questions:

- What is the range of situations (including physical environment, traffic environment, and maintenance conditions) to test autonomous systems?
- Does the concept of "safety margin" apply to autonomous systems based on machine learning?
- How to test high-reliability systems that exploit deep learning algorithms? If a problem surfaces during test, how will the solution be implemented in other systems if the exact state of the system under test is system specific?

High-reliability systems and SoS need to exhibit resilience in the face of internal and external disruptions. Thus, flexible system modeling and deep machine learning play an important role in the system's ability to learn from experience and continue to exhibit increasing levels of resilience. The challenge here is to identify appropriate modeling approaches that have the requisite semantics, are scalable, are amenable to verification and validation, and facilitate test and evaluation.

An AV network is, in essence, a SoS because it exhibits the properties of a SoS (Table 10.2). Therefore, ensuring the safety and reliability of this network is a SoS problem.

10.5 Model-Based Approach to AV-SoS Engineering

Model-based approaches potentially hold high payoff in engineering SoS. However, there are several challenges that need to be addressed before such an approach becomes feasible. These include developing an AV-SoS model that can account for uncertainty, is verifiable, and is amenable to testing; ensuring that the model can handle non-determinism arising from partial observability and uncertainty; ensuring that the model possesses the requisite flexibility to incorporate and exhibit resilience responses to disruptions; and ensuring that the model can scale with an increase in the size of the network and heterogeneity of nodes.

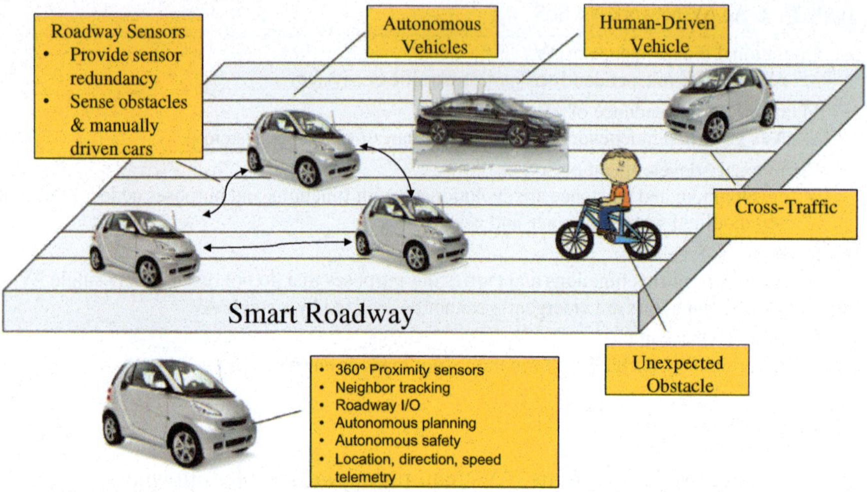

Fig. 10.3 Problem context for resilient AV-SoS design

It is prudent to proceed in stages when addressing these challenges. Initially, certain simplifying assumptions are needed to develop an understanding of the problem. While this problem cuts across several disciplines (e.g., SoS engineering, computer science, law, finance, ethics, and public interaction), the following discussion is confined to the technical problem (i.e., modeling, analysis, and design of a resilient AV-SoS). Specific simplifications in AV-SoS modeling can be made initially in problem framing. These include: (a) having defined lanes for AVs (analogous to high-occupancy vehicle lanes); (b) limiting vehicles in the initial use cases to AVs; (c) specifying AV level of autonomy as SAE level 4; (d) confining safe and resilient behaviors to a defined set of disruptions; and (e) defining the operational environment, disruptions, and variations using a core set of environmental variables. The initial exclusion of human-driven vehicles can be done with the proviso that they are introduced later into the SoS network for the next iteration of modeling, analysis, and design activities. This is because human-driven vehicles are subject to human errors, which are a rich source of disruptions.

With these simplifications, the value proposition of Model-Based Systems Engineering (MBSE) can be explored for AV-SoS networks. MBSE is the formalized application of modeling to support system requirements definition, system design, system analysis, verification and validation, and test and evaluation. It begins with the conceptual design phase and continues through development and into later life cycle phases [12].

In the illustrative AV-SoS example presented in Fig. 10.3, the goal of MBSE is to enable the modeling and analysis of the AV-SoS network with different resilience techniques for various types of disruptions. Figure 10.3 also shows the problem context for exploring and assessing AV-SoS resilience. In this example, networked AVs, equipped with sensors, exploit sensing and networking capabilities to replan,

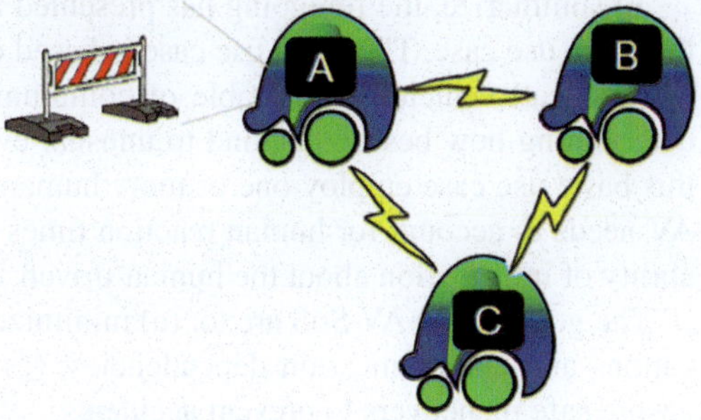

Fig. 10.4 Real-world example: google car rear-ender

reorganize, and adapt to internal and external disruptions. The AVs are capable of interacting with each other and with other entities (e.g., static structures, human-driven cars). The AV-SoS and each AV exploit information acquired from the SoS network in the resilience algorithms to avoid, survive, and learn from disruptions.

In this problem context (Fig. 10.3), self-driving vehicles need to safely negotiate routine hazards, unexpected obstacles, cross-traffic, and human-driven cars. The resilient behavior needed in these vehicles includes the ability to respond to internal and external disruptions at both the AV-SoS and individual AV level. The AV-SoS concept of operations (CONOPS) is as follows. An AV receives inputs from a variety of sensors on stationary objects and other vehicles. The AV-SoS uses this information to adapt behavior safely (i.e., exhibit resilience) at the SoS level as well as at the individual AV level. A real-world example of the Google car illustrates these concepts (Fig. 10.4).

Real-World Example The Google car rear-ender early in 2016 provides an excellent illustration of a real-world example. The Google AV was rear-ended by a bus when it changed lanes to avoid sandbags but failed to account for the bus. In a fully autonomous SoS network, the vehicle would have been able to coordinate actions and avoid this collision. This example provides the necessary information to define an exemplar use case and alternative use cases.

Examplar Use Case All vehicles in the SoS network (compromising vehicles A, B, and C) are autonomous vehicles (i.e., SAE level 4 or 5). Vehicle A senses an obstacle and wants to move into the lane occupied by Vehicle C. Vehicle A communicates its intent to Vehicles B and C, to verify safety of proposed lane change. If Vehicles B and C confirm that lane change is safe, Vehicle A performs the lane change. If lane change is unsafe, then Vehicle A applies brakes.

Alternate Use Case #1 Vehicle C is not autonomous. In this case, Vehicle A signals Vehicle B and proceeds to apply brakes.

Alternate Use Case #2 Neither Vehicle B nor Vehicle C is autonomous. Vehicle A updates its speed based on proximity of Vehicles B and C. Then Vehicle A may only cross into Vehicle C's lane, if sufficient, reliable information about Vehicle C's location and speed is available.

To summarize, the foregoing has presented a basic use case and two variants of the basic use case. The basic use case is based on a fully autonomous SoS network in which all vehicles are capable of communicating with each other and jointly determining how best to respond to unusual events or disruptions. The variants of this basic use case employ one or more human-driven vehicles. In these cases, the AV needs to account for human reaction times as well as location, speed, and reliability of information about the human-driven vehicles in its decision making.

The goals of the AV-SoS are to: (a) minimize accidents; (b) minimize communications and communication dependencies; (c) maximize flow efficiency; and (d) ensure safe maneuvers to prevent accidents.

With respect to SoS resilience, AVs will employ different parameters and/or rules in resilience algorithms than those used in automated, human-driven vehicles. As important, the AV-SoS will be capable of determining which use case applies and makes changes to parameters accordingly.

Introducing resilience in the SoS network requires a set of use cases. Once again, for simplicity, let us assume a fully autonomous SoS in which all vehicles are AVs. Let us also assume that AV-SoS planning is hierarchical and the communication protocols followed by the AVs are preplanned, i.e., vehicles follow preplanned rules, or are *context-driven*, i.e., a limited set of AVs are correlated based on a use case pattern.

Preplanned behaviors are analogous to *mission system tests* (MST) performed for spacecraft missions, while context-driven behaviors are similar to *operational readiness tests* (ORTs) that are also performed for spacecraft missions. MSTs are performed when we know the answer and we are interested in verifying that the SoS does what is expected. ORTs are performed when we do not know the answer but are interested in ascertaining that the SoS remains safe.

10.6 AV-SoS Use Cases and Behavior Patterns

Defining use cases for AV-SoS is challenging because there are numerous scenarios to choose from. However, several scenarios do not correlate well with today's driving scenarios. Therefore, the choice of scenarios should be based on how well they help convey resilience characteristics and how well they enable AV-SoS testing. Complicating factors, associated with the heterogeneous environment include *legal* (e.g., making right turn on red when it is not allowed, jaywalking, street parking, time windows for laws, e.g., HOV lane usage, parking); *informational* (e.g., presence of signs, color, and type of signs or signals, marking of emergency vehicles, crosswalk marking); and *maintenance of infrastructure and neighboring systems* (e.g., faded road lines, graffiti on signs; broken lights, poorly maintained human-driven cars on the road). And finally choosing the autonomy level (from SAE-defined autonomy levels) impacts SoS behavior complexity. By choosing SAE level 4, we can avoid the complex issues associated with the presence of humans in vehicle control.

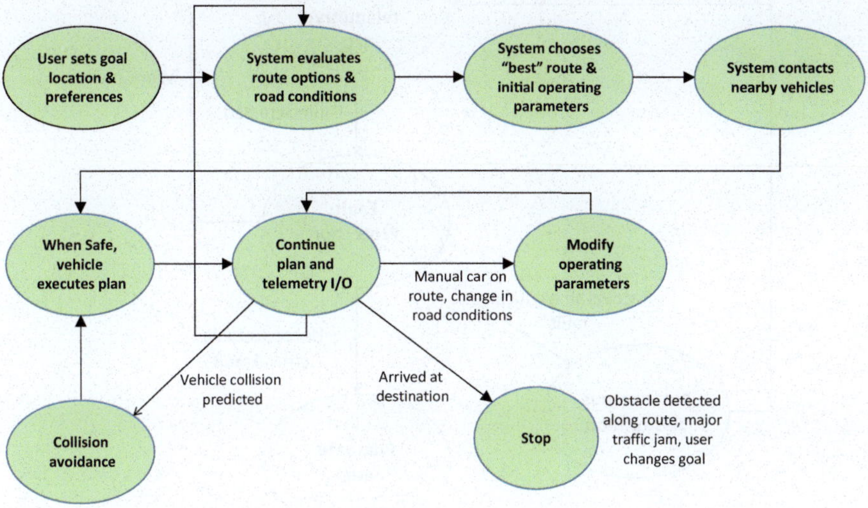

Fig. 10.5 Nominal (predictable) AV operation

Against the foregoing backdrop, we can model nominal and off-nominal (predictable) AV operations (Figs. 10.5 and 10.6).

AV Control The control architecture of an AV is shown in Fig. 10.7.

The vehicle control plan comprises location-dependent commands that are sent to the vehicle controller. The vehicle controller accepts state estimates and updates configuration goals which are sent to the deductive controller. The deductive controller accepts configuration goals (from the vehicle controller), inputs from the vehicle model, observations (from the vehicles), environmental inputs (from the environment sensors), to produce state estimates for the vehicle controller, and vehicle command which are sent to the vehicle to achieve a desired interim/end goal.

Use Cases Use cases for AVs are associated with two types of patterns: those experienced by human-driven cars and those that apply exclusively to AVs. Table 10.3 presents an overview of AV behavior patterns that can be used to develop AV use cases.

The following paragraphs discuss these AV behavior patterns:

1. Highway merge (on-ramp or off-ramp)

   Merging is a coming together or blending of vehicles to maintain a smooth flow of traffic. The main challenge in performing this behavior is sensory in that the merging vehicle cannot always see oncoming traffic. In human-driven cars, humans "judge" relative speeds to merge. Introducing this capability in self-driving cars is a challenge. A further complication is that merge difficulty varies as a function of traffic and road/type quality. Merge behavior tests fore, aft, and lateral car control and sensing coupled with dynamic uncertainty.

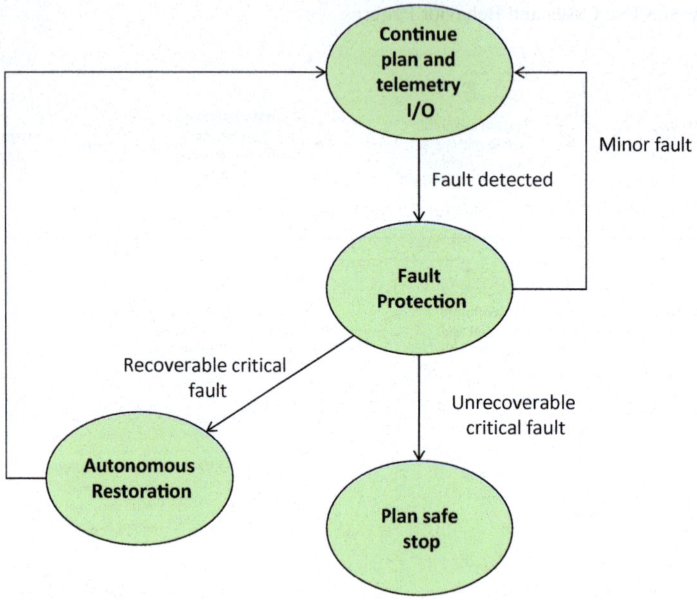

Fig. 10.6 Off-nominal (predictable) autonomous vehicle operation

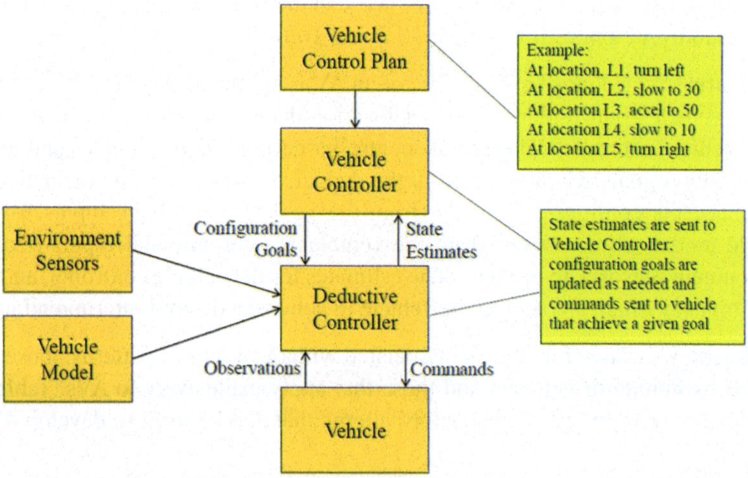

Fig. 10.7 AV control architecture

Table 10.3 Pattern-driven AV use cases

• Highway merge (on-ramp or off-ramp)
• Four-way stop, heterogeneous systems
• Lane blockage
• Unprotected left turn
• Highway driving
• Rural/country road driving
• City driving (environmental density)

2. Four-way stop

   A four-way stop is associated with a four-way intersection of two-way roads controlled by stop signs or road markings. This context is different from stoplights where lights protect certain maneuvers. A four-way stop requires negotiation between self and other vehicles in accord with "right-of-way" rules. The problem is relatively simple when all vehicles are autonomous. However, when the vehicles in question are a mix of AVs and human-driven cars, the problem becomes challenging. The four-way stop tests dynamic decision making in uncertain, heterogeneous (i.e., AV and human-driven vehicles) vehicle environments. Human drivers rarely follow the "letter of the law" in these cases but still manage to negotiate such challenging situations safely.
3. Lane blockage

   This is a common occurrence on a two-way road. Examples of static blockage are disabled vehicle, felled tree, construction, and weather effects (e.g., flash flooding). Examples of dynamic blockages/obstacles include traffic accident, pedestrian, animal, or child that darts into the lane. Successful execution of an avoidance maneuver requires environmental reasoning with respect to rules of the road and new developments that require an avoidance or stopping maneuver and adaptive decision making such as the AV deciding to cross the yellow line for a brief moment using sensory data to perform a maneuver outside the norm.
4. Highway cruise

   This pattern is associated with multilane highway driving. It primarily tests navigational and vehicular controls. This maneuver is relatively straightforward in comparison to others. Even in dense environments, objective variance among vehicles (agents) is quite small due to road type restrictions. This pattern can serve as the baseline scenario for introducing "injects" such as hazards, weather changes, merging, scaling traffic, and responding to vehicle emergency. This maneuver requires environmental awareness of speed – which is the test for this pattern.
5. Unprotected left turn

   This pattern is associated with a T-intersection in which the perpendicular road does not have a stoplight to ensure protected left turns. The challenge tests that the AV needs to meet are judging openings based on multisensory data, assessing risk, and understanding and predicting the behaviors of human (agent)-driven vehicles. This pattern can be used for scenario injects (accident avoidance, how to handle a looming accident).
6. Rural/country road cruise

   This pattern, which is associated with a two-way road, is generally more difficult and riskier than highway cruise. This is because physical lines of sight vary as do road quality and type. The tests associated with this pattern that an AV needs to pass are environmental awareness at speed, sensing in diverse and dense environments, and communication capabilities. This pattern serves as a baseline to inject other events similar to those for the highway cruise pattern (use case).

7. City driving

   This pattern is associated with driving in dense areas comprising automobiles, infrastructures, pedestrians, motorcyclists/cyclists, traffic police officers, construction, trains, buses, and roadside parking. This pattern is also associated with dynamic environments in which other than infrastructure, most elements move and involve humans (agents). The third characteristic is fluctuating speeds, i.e., transition from slow to fast and back to slow. The challenges confronting this behavioral pattern are data handling and sensor capabilities in data-rich environments and, specifically, partitioning data by importance, making local decisions in light of holistic objective and information picture.

Use Case Modifiers ("injects"). Several factors can serve as modifiers in use cases. One of the most important modifiers is "weather." Inclement weather affects several variables used to define use cases. These include sensor capability, vehicle control capability, route planning (avoidance, detour, …), environmental awareness, decreased performance of neighboring vehicles (agents), and AV decision making (navigational, abort mission criteria, uncertainty). A study by Wachenfeld et al. (2016) defines success for driving in weather as "The quality as well as the success rate with which the driving robot performs the driving task is similar to the human quality and success rate." [13]. For example, ceasing operation due to inclement weather occurs only when a human driver would also discontinue the journey as well [13].

Other "injects" into use cases include yielding to emergency vehicles, parking, and pulling over. Yielding to emergency vehicles tests sensor capabilities and communication capabilities (long haul and across the ad hoc mesh) as well as environmental reasoning while counterintuitively maneuvering the vehicle (e.g., pulling off road on unpaved shoulder to let emergency vehicle pass). Parking in a garage, roadside or lot can be its own use case or added as a pre-use case or post-use case (i.e., additional activity) to an existing use case (pattern). This use case tests environmental deduction and induction about the state of vehicles – which vehicles are parked, which are standing, and which are waiting for a parking spot. Pulling-over use case is associated with a mechanical malfunction or fault or other emergency/unusual situation.

10.7 Formal Modeling of AV-SoS

Formal methods have been in use in the chip design industry for decades. Today formal methods are being used to model both cyber-physical-social systems and SoS. Their key distinguishing characteristic is that they employ mathematically rigorous semantics. This property ensures high degree of confidence in the analyses performed with such models because model correctness can be verified. Model correctness implies model completeness, consistency, and traceability. As important, formal methods facilitate model (i.e., system) testing. The set of questions that the model is expected to answer defines the intended purpose of the model. As important, the application of formal methods tends to reduce the risks in the development process by reducing the likelihood of late detection of defects, as well as reducing the cost of defect detection.

Static analysis approaches enabled by formal modeling include model checking and formal theorem proving. These methods when applied to models created in upfront engineering help identify areas of incompleteness and ambiguity in system or SoS requirements and specifications. When applied to later stages in design or code-level analysis, they can be used to identify specific patterns of defect. Formal methods enable system developers to mathematically prove that the system or SoS models developed exhibit essential properties such as resilience and security.

Formal methods are important for AV-SoS model verification and testing. Since a SoS is a complex software-intensive system (SIS), it is not possible to prove that it is error-free by merely testing it. This is because testing every possible path through the software code, for every combination of data that could potentially cause that path to fail, and then checking that each path led to a correct result, is impractical. It would take an inordinate amount of time and incur prohibitive costs, even if it were possible. This is in sharp contrast to physical systems (e.g., aircraft wing) which obey the laws of physics (i.e., natural laws) that allow engineers to test a design at extreme values and infer system behavior between the extremes [14]. In a software-intensive SoS, the behavior of the SoS depends largely on software logic and not physical laws (note: the behavior of an individual AV does depend on physical laws). Thus, for a SoS, it cannot be assumed that because the software logic functions as intended for some input values, it will function as intended for other input values, even if the latter lie between the extremes defined by the former.

The implication of the above is that SIS invariably contains latent errors, even after rigorous testing during development. Studies within academia and industry have shown that there are typically 5–30 errors in 1000 lines of code at the time that the SIS goes into service, following normal testing. This is where the value proposition of formal methods comes into play.

With the use of formal methods even for a part of the development process, it becomes possible for software tools to analyze models and detect errors more quickly, inexpensively, and comprehensively than with conventional testing methods. Formal methods are capable of supporting automated testing (i.e., test generation, test execution, test output checking). This, in essence, is "model-based testing" with the added rigor provided by formal semantics. In the software world, formal model-based testing has shown significant time and cost savings when compared to traditional manual testing.

Modeling an AV and an AV network is the starting point for the design of the AV-SoS. There are several requirements that the selected modeling constructs need to satisfy: enable *verification of correctness* (i.e., completeness, consistency, and traceability), ensure *scalability* with the number and heterogeneity of AVs in the network, support *test and evaluation*, and ensure *satisfaction of desired design quality attributes*.

For a SoS, several formal methods can be employed that range from deterministic to probabilistic. Examples of deterministic methods are computational tree logic (CTL), linear temporal logic (LTL), and contract-based design (CBD). Examples of probabilistic or stochastic methods are Hidden Markov Models (HMM) and Partially Observable Markov Decision Processes (POMDPs). Deterministic models are useful in verification and testing, while probabilistic models are helpful in dealing with partial observability of the SoS states and uncertainty in the environments.

With the above requirements in mind, a modeling approach that is a combination of deterministic and probabilistic approaches can be employed. Specifically, a combination of CBD and POMDP can be employed. CBD is a formal method for explicitly defining, verifying, and validating system requirements, constraints, and interfaces. With CBD, an implementation satisfies a design constraint if it fulfills guarantees when assertions are true. In the CBD approach, design is a composition process, in which the system design is the result of successive assembly of components. A component contract specifies assumptions about its environment and guarantees about its behavior. A composition is said to be "well formed" if the assumptions of each component are contained in guarantees offered by others. The appeal of CBD stems from the fact that statements in a contract are mathematically provable. The limitation of CBD is that the assertions are invariant. From a resilience perspective, it becomes necessary to relax the requirement for invariant assertions to introduce flexibility. Flexibility is introduced through the use of POMDP.

The rationale for choosing POMDP stems from the fact that even though many real-world complex problems and systems are only partially observable, the Markov assumption invariably holds. A POMDP model comprises a set of states S, a set of actions A, and a set of observations O. A POMDP comprises a transition model, a reward model, and an observation model. The Markov assumptions apply to the transition model, with the optimal policy depending only on the current state. POMDP is an effective modeling approach when the operational environment is only partially observable, with the current state usually not known. Therefore, the agent (i.e., autonomous system) cannot execute the optimal policy for that state. In the suggested approach, contracts become flexible through relaxation of time-invariance restriction on the state space and action space, adding an evaluation metric to POMDP to determine best action, updating emission and transition probabilities of the hidden state, and finally adding the concept of time to POMDP. Then, the CBD without time-invariance restriction is augmented with POMDP with the above additions. We call the resultant contract, which extends the deterministic contract representation to stochastic systems, a resilience contract (RC). The key features of the resilience contract are in-use learning, uncertainty handling, and pattern recognition. A resilience contract, which is specified during system design, is "trained" when the system is put into actual use ("learning").

10.8 Lingering Misconceptions About AVs

A few misconceptions about AVs still remain even as AVs are beginning to make their way into controlled operational environments. These include:

Misconception 1: AVs Will Evolve Naturally from Automated Vehicles
Features such as function automation (e.g., auto-park, lane warning, adaptive cruise control, emergency braking) do not extrapolate to an AV because they are designed for human-driven vehicles and are architected based on human-centered design principles – "machine-aided man" paradigm [15]. With the human in a central role in this paradigm, human cognitive limitations, transparent vehicle behavior, and human override and intervention points need to be explicitly addressed [4]. On the other hand, AVs are architected and optimized without regard to a human driver or supervisor. Rather, they are architected for performance and specific nonfunctional requirements such as dependability and resilience. As important, driver-assist function automation in automated vehicles is designed for short-term operation in limited, structured settings. For example, auto-parking operates for but a few seconds under the watchful eye of the driver. The same is true of *emergency braking* which is triggered just prior to an imminent crash and *lane warning alerts* which are triggered when the vehicle wanders outside its lane. These features, as implemented in automated vehicles, have little value for AVs, which need to respond to all eventualities that can occur during extended vehicle operation. Features associated with supervised function automation that address structured tasks are inherently incompatible with this need. The bottom line is that a system designed to serve the needs of human drivers cannot morph into an AV because the assumptions, architecture, and design of human-driven vehicles with selective function automation are quite different from those for AVs.

Misconception 2: AVs Will Employ Classical Control Algorithms
AVs are not preprogrammed to execute some monolithic control algorithm. Rather, they "learn" from a large number of traffic situations. Specifically, AVs rely on smart sensing, deep machine learning, and pattern recognition techniques. Deep machine learning, a form of artificial intelligence (AI), provides computers with the ability to learn from observations without requiring explicit programming. For example, the Google car can adjust on the fly using its machine learning capability. Using data analytics, it can learn to react to a vehicle that cuts into its lane or a pedestrian that, on impulse, decides to dart across the street right in front of it. AVs are able to collect, process, store, and recall this type of data in real time. To distinguish one object from another, a sophisticated learning algorithm is trained using multiple representative images containing various objects of interest. Each image is annotated with the particulars of the object it contains. The machine learning algorithm incrementally learns from these images, initially making pure guesses (mostly wrong) and incrementally modifying its parameters and/or structure until it reaches an acceptable classification accuracy even with new, previously unseen images containing the objects of interest. This learning approach can also be used for actions

and evaluations. In the latter case, the software developer feeds the software with a variety of representative traffic situations and specifies the correct answer for each. The learning software searches for the best values of internal parameters and best configuration of decision logic that allows it to act correctly in various situations. For example, Google continues to assemble an extensive library of traffic situations from which its AVs can learn and become smarter. The probabilistic learning algorithm is another approach employed by AVs. For example, an AV does not assume it knows its exact position. Instead, it maintains a probability distribution of its current position and chooses the position with the highest probability. Thus, AVs are autonomous systems that continually learn from experience.

Misconception 3: AVs Are Expected to Make the Right Ethical Decisions
A networked AV equipped with sensors and reasoning can be expected to identify nearly all decisions with ethical implications. Routine ethical decisions are already being addressed today by AVs. For example, Google's AV is being equipped with reasoning facilities for debris avoidance [9]. On rare occasions, a child or an animal might dart into the street in front of an AV. Such incidents pose risks, and ethical decision making is needed when such risks exist. And risks do exist in driving, regardless of whether the AV is on an open highway, or on a crowded street. A majority of these ethical problems are straightforward and can be readily resolved by AVs. On exceedingly rare occasions, an AV could find itself in an ethical dilemma for which no good options exist. The likelihood of an AV facing such a "no-win" situation is negligible. Today billions of commutes are made daily with several thousand people dying each day in traffic accidents. The incidence of ethical dilemmas in these casualties is miniscule. Therefore, practically speaking, while such extreme hypotheticals may provide fodder for debate, they are largely irrelevant for AVs. The incidence of even extremely rare ethical dilemmas can be further minimized by exercising proper control of the environment (e.g., assigning a dedicated lane to AVs) and through government regulations.

Misconception 4: AVs Will Be More Environmentally Friendly
Not necessarily. It depends on how technologies advance and the degree to which the environment can be controlled. Since AVs will be less erratic than human-driven cars, they will use less fuel. On the flip side, it is likely that there will be many more of them on the streets for a variety of reasons (e.g., more efficient spacing on streets and highways). Furthermore, the use of unoccupied AVs to pick up and drop off passengers, and allow family members use of the vehicle, is likely to increase total miles driven. As important, when AVs operating within a SoS network are GPS-equipped, intelligent routing algorithms can be employed to dynamically allocate traffic flows to avoid residential neighborhoods and thereby reduce pollution in residential areas.

Misconception 5: AVs Will Alleviate Traffic Congestion
GPS-equipped AVs within an AV-SoS network will be able to signal each other about congestion, accidents, and vehicle breakdowns ahead of time to avoid traffic jams and bottlenecks. They could also employ intelligent routing algorithms to keep

traffic flowing on parallel paths and thereby reduce traffic congestion and commute times. At the same time, with these improvements in place, more vehicles can be expected to be on highways and streets, thereby creating congestion. This phenomenon has shades of what is termed a "wicked" problem in which the remedy to a problem causes the problem to resurface based on how change propagates through the complex system.

Misconception 6: AVs Will Be Totally Safe

This assumption, which implies that the act of self-driving poses no risk, is clearly erroneous in that it claims complete safety regardless of the environmental factors at play and the fallibility of subsystems within an AV. Unfortunately, subsystems can fail, and human-driven vehicles, pedestrians, and motorcyclists/bicyclists will always be sources of potential accidents. Goodall asserts that crashes are possible with imperfect systems, perfect systems with driverless and human-driven vehicles in traffic, and perfect systems without human-driven vehicles in traffic [9].

To confirm the veracity of his claim, let us look at the extreme case of only AVs on the streets and highways. Even in this extreme case, accidents can happen because the subsystems in AVs can fail, and latent software glitches can incapacitate vehicles and cause accidents. And, of course, once AVs begin to use the same streets as human-driven vehicles, the likelihood of collisions increases because of the ultra-sensitivity of AVs to unexpected incursions on the street.

With AVs, as with human-driven vehicles, the likelihood of encountering debris, animals, and inclement weather exists. And, of course, there is always the very real danger of failures in the hardware, software, and perceptual and reasoning facilities of AVs. Any of these factors, or combination of factors, can result in collisions among AVs.

To ensure its own safety, an AV needs to maintain both situation awareness and risk awareness. This means maintaining cognizance of speeds when negotiating curves, and being more vigilant when momentarily crossing the centerline to pass a cyclist, or deliberately violating traffic regulations to avoid dangerous situations that suddenly arise due to distracted drivers, unexpected obstacles, and the like. For example, an AV needs to decide instantaneously how best to allocate risks between itself, adjacent vehicles, people, and built-up structures. These time-stressed decisions have to be made in real time by the AV to avoid accidents where possible or minimize damage to affected parties when a crash is imminent and unavoidable.

Misconception 7: An AV that Follows Traffic Regulations and Laws Will Automatically Satisfy Ethics

First and foremost, existing laws do not cover every eventuality and are not specific enough to realize reasonable computer responses. A simple example is that of debris in a lane [9]. A human driver would instinctively cross the lane demarcation line to avoid the debris in the absence of oncoming traffic [9]. However, an AV designed to follow the law literally would stop and wait for the debris to be cleared – an impractical proposition [9]. While in theory laws can be added to deal with such situations, the problem quickly becomes exponentially difficult. For example, how does one define obstruction and safe [9]? How does one make sure that such terms are computer-understandable? Such questions are being addressed today for AVs.

Misconception 8: Reduction in Fatalities Is the Key Metric
Autonomy can be expected to eventually reduce fatalities from automobile accidents. However, it can be argued that reduction in overall fatalities is the wrong metric when viewed through the lens of ethics [16, 17]. This is because improved safety for one group may well come at the expense of another. For example, if vehicle fatalities drop but bicyclist fatalities increase despite reduction in overall traffic fatality rates, society can be expected to take a dim view of such outcomes. This assumption is also problematic because it assumes that the primary objective of an AV is improving safety. While certainly an objective, there are other objectives for AVs. These include mobility for those unable to drive, reducing transportation costs (and therefore the costs of goods) and reclaiming urban areas by optimizing traffic flow. Recall Virilio's refrain that the invention of the airplane invented the airplane crash [6]. The pursuit of societal advancement is not without risks, and while every life lost in an airplane crash is unacceptable, air transportation has forever changed the world in positive ways. Thus, while maximizing lives saved is a noble and important objective for AVs, making it the sole metric represents a narrowly focused utilitarian view. It may also be ineffective in that the value proposition for society is highly context dependent. As Goodall notes, "[R]isk of death from nuclear meltdown is often over-valued, while traffic fatalities are undervalued" [9]. Therefore, the challenge is to accurately reflect societal values across a wide and representative range of routine driving and accident scenarios and then translate them into a machine-understandable language.

10.9 Liability Management

The liability issue with AVs is significantly more complex than with human-driven vehicles. When human-driven automobiles get into an accident, the personal insurance of the party at fault typically covers the costs of damage to the victim's vehicle and medical bills for injury sustained by the victim. However, with AVs, there is no straightforward way to allocate liability among the different parties. This issue needs to be resolved before AVs are deployed in large quantities on our roads, streets, and highways.

There are other related questions that need to be answered before addressing how to distribute liability:

- Beyond adherence to the law, what else does an AV have to do?
- Should AV software be flexible enough to break the law for a greater good?
- How to ensure that programmers are not placed in the unenviable role of having to decide what is safe versus what is legal?
- Does an AV ensure the safety of its occupants over all others (i.e., other vehicles, pedestrians)? If so, should the AV be programmed to be selfish, to ensure its occupants' safety first?
- Who all are liable/legally responsible for accidents that occur with AVs? Car owners? Manufacturers? Programmers? Regulators? All? If all, how should the liability be distributed?

- Are there any specific circumstances, in which AVs should be allowed to break the law?
- Who all should contribute to answering these ethical questions? Should the answers be determined collaboratively by the automotive industry, policy makers, attorneys, and philosophers?
- Can legality be traded for safety in certain contexts?

A Business Insider Intelligence report forecasts that as many as ten million AVs will be in operation by 2020. Assuming this forecast is reliable, there is precious little time to resolve the liability issue. According to Mayerson, "in theory, AVs would not create negligence liability for the passenger/non-driver/car owner." Today, for most drivers, liability coverage accounts for a large part of their automobile insurance premiums. The cost of repairing physical damage to a vehicle is but a fraction of costs of hospital bills and court fees paid to victims. However, if liability is taken out of the equation, personal automobile insurance premiums for owners of AVs would be substantially lower than they are today. This begs the question: who should be held accountable for liability costs? One model calls for the automobile manufacturer to bear all the liability and impose that liability simply based on the autonomous car being a substantial cause of the injury. In other words, if the accident is the result of computer failure, it is reasonable to assign blame to the vehicle's manufacturer. However, there are unintended consequences that can ensue from such a law. Passing on insurance costs to automakers is not necessarily in the best interests of consumers because automakers, confronted with increasing operating costs due to liability concerns, will attempt to recoup those costs by increasing vehicle prices. In other words, owners of AVs may pay a smaller premium but at the expense of a larger price tag for AVs!

Modeling and simulation approaches currently used in the engineering of today's automobiles can be leveraged for V&V of AVs. However, it is not possible to envision and perform V&V for every possible scenario instance. Therefore, at what point does the V&V performed on the AV, adherence to laws and regulations, and proper use of modeling and simulation absolve a manufacturer from potential liability? The answer to this question as framed is not clear-cut. An effective approach to address this problem is to map scenario instances to scenario patterns and conduct V&V using lightweight formal methods on those patterns with the assumption that V&V of patterns is tantamount to covering the scenario instances that define the operational space.

10.10 Summary and Outlook for the Future

Advances in AVs and AV-SoS are being enabled by hyper-connectivity and convergence. Hyper-connectivity has enabled AVs to communicate with human-driven vehicles and smart structures. Convergence has taken many different forms. The first type of convergence is the adaptation and integration of deterministic and probabilistic methods to create new AV and AV-SoS models that on one hand are verifiable and testable and on the other process sufficient flexibility for responding in resilient fashion to disruptions. The second type of convergence is the exploitation of the Internet of Things (IoT) in creating an AV-SoS network. The third type of convergence is making "dumb" infrastructures smart through embedded sensors, so they can be incorporated into the AV-SoS network. The fourth type of convergence is embedding security in software during software development. There are other types of convergence such as convergence of engineering with decision making and ethics, and the convergence of business and engineering to create new business models that make AV-SoS networks a commercial success.

According to the Gartner Group "Hype Curve," with any new technology, there is a peak in hype, followed by a trough of disillusionments. While assessing the maturity level of AVs, Gerdes suggested that "we are somewhere on that hype peak at the moment." He cautions, "The benefits are real, but we may have a valley ahead of us before we see all of the society-transforming benefits of this sort of technology."

AVs will undoubtedly have an impact on the economy. According to the McKinsey report, the economic impact will be felt in stages with increasing impacts along the way. The report describes three distinct phases.

Current to 2020 During this period, AVs will be well incorporated into industrial, controlled environments such as farms. At the same time, passenger vehicles will be going through a relatively long prototyping and testing period. Along the way, autonomous aspects will be introduced and tested in traditional automobile models. However, commercially available AVs are not expected to appear until 2020. According to McKinsey, some economic benefits in the form of industrial efficiency will be realized, with the majority of economic benefits yet to come.

2020 to 2040 During this period, AVs will become mainstream. The technology can be expected to stabilize from initial feedback. It is expected that vehicle design will change with several features (e.g., gas pedal, brake) becoming redundant. Consumers will have discretionary time at their disposal during commutes. Technology costs will continue to fall as economies of scale come into play, and suppliers rush to adapt to new configuration and functionalities of AVs. During this period, several industries are likely to be impacted. Cab drivers, delivery vans, and truckers are likely to become increasingly redundant. The automobile insurance business will be redefined as liability shifts from drivers to automobile manufacturers. As important, automobile repair shops will require different expertise and will be fewer in number once remote repairs become feasible.

2040 and Beyond During this period, AVs will become commonplace raising the question: Will individually owned vehicles still be needed for transport? While automobile manufacturers may insist that they will be needed, the reality is that 95% of the time most vehicles go unused. It is entirely likely that ride-sharing and shuttling services will become increasingly popular. Furthermore, according to McKinsey, since AVs are capable of self-parking, parking areas will be used more efficiently, freeing up approximately 25% parking space. And, of course, other professions are also likely to be impacted by new rules of transport. For example, with fewer traffic-related accidents and injuries, it will become possible to reassign police personnel to other law enforcement and security duties. McKinsey envisions a safer environment with 90% decrease in accidents resulting in $180 billion in savings in repair and healthcare costs alone. The McKinsey report suggests that the overall economic potential from AV technology will be between $0.2 trillion and $1.9 trillion annually by 2025. According to the Eno Center for Transportation study, with 90% AVs in the United States, as many as 4.2 million accidents will be avoided, saving 21,700 lives and $450B in related costs. Reduced stress, from shorter commutes to work in crowded metropolitan areas, is another significant health benefit of AVs.

One can expect a clash of business models between incumbents and new players in the AV space over the next two decades. This clash will disrupt not only their existing business models but also the business models of healthcare providers, insurers, dealers, suppliers, customers, regulatory bodies, and legal entities. It is unclear how this interdependent future will unfold over time. However, what is clear is that AV-SoS networks are inevitable in the future, so we need to prepare for them.

Ultimately, it all comes down to what is "safe," "legal," and "ethical" and who all are "liable." What is safe may not always be legal. What is legal may not be ethical. What is ethical may not be safe to certain parties. Who is liable in many cases may not be clear. A lot of ground needs to be covered before we ask and answer such questions. To those who believe that the technology is here, it is worth reminding them that it takes more than technology to realize a viable AV. To those who believe in utilitarianism, it is worth reminding that collective benefit of a group trumps individual benefit unless that individual happens to be one of them! To those who believe in deontological ethics, it is worth reminding them that humans are capable of common sense reasoning while today's computers tend to literally interpret statements and commands. The latter can give rise to unexpected, unpredictable behaviors, thereby jeopardizing safety. To those who believe that the passengers of AVs need to be served first regardless of collateral damage, it is worth reminding them that this view could be a violation of ethics. Upon further reflection, it should become clear that we are not close to where we want to be, but we are a far cry from where we started. At least we are beginning to ask the right questions. That's a start.

References

1. Viereckl R, Ahlemann D, Koster A, Hirsh E, Kuhnert F, Mohs J, et al. Connected car report 2016: Opportunities, risk, and turmoil on the road to autonomous vehicles 2016 [Available from: http://www.strategyand.pwc.com/reports/connected-car-2016-study.
2. Kortuem G, Kawsar F, Sundramoorthy V, Fitton D. Smart objects as building blocks for the internet of things. IEEE Internet Comput. 2010;14(1):44–51.
3. Madni AM, Sievers M. System of systems integration: key considerations and challenges. Syst Eng. 2013;17(3):330–47.
4. Madni AM. Integrating humans with software and systems: technical challenges and a research agenda. Syst Eng. 2010;13(3):232–45.

5. Virilio P. The original accident. Cambridge: Polity; 2007.
6. Chatfield T. Automated ethics: can we design machines to make ethical decisions? . 2015.
7. Wired. Google's self-driving car caused its first crash 2016 [Available from: http://www.wired.com/2016/02/googles-self-driving-car-may-caused-first-crash/.
8. Review MT. Why self-driving cars must be programmed to kill 2015 [Available from: https://www.technologyreview.com/s/542626/why-self-driving-cars-must-be-programmed-to-kill/
9. Goodall NJ. Machine ethics and automated vehicles. In: Road vehicle automation: Springer Science + Business Media; New York, NY, 2014. p. 93–102.
10. Bonnefon J-F, Shariff A, Rahwan I. Autonomous vehicles need experimental ethics: are we ready for utilitarian cars? arXiv preprint arXiv:151003346. 2015.
11. Garza AP. Look ma, no hands: wrinkles and wrecks in the age of autonomous vehicles. New Eng Rev. 2011;46:581.
12. INCOSE. Systems Engineering Vision 2020. INCOSE-TP-2004-004-02). http://www.incose.org/ProductsPubs/pdf/SEVision2020_20071003_v2_03.pdf, 2007.
13. Wachenfeld W, Winner H, Gerdes JC, Lenz B, Maurer M, Beiker S, et al. Use cases for autonomous driving: Autonomous Driving. Germany: Springer; 2016. p. 9–37.
14. The Institute of Engineering and Technology, Formal Methods: A Factfile 2011.
15. Madni A. The role of human factors in expert systems design and acceptance. Hum Factors J Hum Factors Ergon Soc. 1988;30(4):395–414.
16. Science I. When driver error becomes programming error 2016 [Available from: https://www.insidescience.org/content/when-driver-error-becomes-programming-error/2591.
17. Lin P, Abney K, Bekey GA. Robot ethics: the ethical and social implications of robotics. Cambridge: MIT press; 2011.

Chapter 11
Looking to the Future

"The illiterate of the 21st century will not be those who cannot read and write, but those who cannot learn, unlearn, and relearn."

– Alvin Toffler, writer and futurist

11.1 Promise of Transdisciplinary Systems Engineering

In 2008, the US NAE identified 14 "grand challenges for engineering" for the twenty-first century. These grand challenges shown in Table 11.1 can be conveniently grouped into four categories: sustainability, health, vulnerability, and joy of living. What is common to these challenges is that they are all complex systems problems, and they all require contributions from multiple disciplines to frame the problem correctly. For example, *restore and improve the urban infrastructure* requires expertise in construction engineering, infrastructure engineering, information technology, sociology, culture, and resilient systems engineering. Similarly, *reverse-engineering the brain* requires expertise in neurocognitive science, computer modeling, and systems engineering. *Secure cyberspace* requires methods from psychology/AI (adversary modeling), biology (architectures), computer science, and cybersecurity systems engineering. "Advance personalized learning" requires methods from learning sciences, human-technology integration, machine learning, and multimedia/multimodal presentation of learning content, and human-system integration engineering. Reaching the goals of each one of these thrusts is being made possible through growing disciplinary convergence — the key enabler of transdisciplinary systems engineering. In the 21st century, we can expect to see increasing attention on transdisciplinary systems engineering as we continue to pursue solutions to these grand challenges problems.

The previous ten chapters have presented twenty-first century trends that will continue to impact how systems will be engineered in the future. On the one hand, we have to contend with hyper-connectivity and increasing system complexity. On the other hand, we have the opportunity to exploit disruptive technology advances and disciplinary convergence. The former poses challenges. The latter affords opportunities for innovation to respond to these challenges. The need for new thinking to exploit disciplinary convergence has been emphasized along with

A.M. Madni, *Transdisciplinary Systems Engineering*,
DOI 10.1007/978-3-319-62184-5_11

Table 11.1 NAE grand challenges

Sustainability • Make solar energy economical • Provide energy from fusion • Develop carbon sequestration methods • Manage the nitrogen cycle • Provide access to clean water • Restore and improve urban infrastructure
Vulnerability • Prevent nuclear terror • Secure cyberspace
Health • Advance health informatics • Engineer better medicines • Reverse-engineer the brain
Joy of Living • Enhance virtual reality • Advance personalized learning • Engineer the tools of scientific discovery

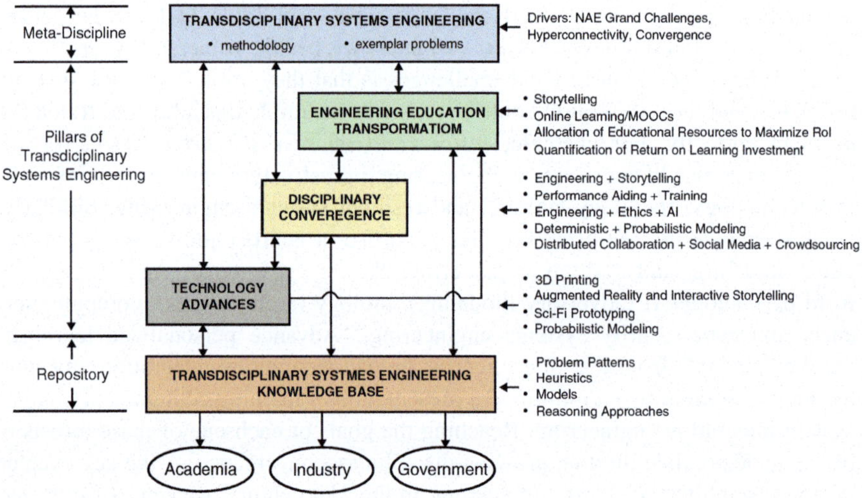

Fig. 11.1 Evolution of transdisciplinary systems engineering

specific types of new thinking that encourages transdisciplinary thinking. Specific examples of disciplinary convergence were presented to illustrate the creation of new system capabilities. Collectively, these new concepts and new types of thinking constitute an initial body of knowledge for advancing transdisciplinary systems engineering.

Figure 11.1 presents a conceptual framework for evolving new concepts and content in support of transdisciplinary systems engineering. As shown in this figure, the three enablers of transdisciplinary systems engineering are disruptive technology advances, ongoing disciplinary convergence, and ongoing transformation in engineering education. Collectively, advances made in these three areas will contribute to the evolving body of knowledge in transdisciplinary systems engineering.

Successful adoption of transdisciplinary systems engineering requires diffusion of transdisciplinary SE concepts, principles, and perspectives in different sectors so they become part of our societal fabric. To achieve successful diffusion, the value proposition of transdisciplinary systems engineering has to be proven in multiple domains (e.g., defense, healthcare, energy, education). In particular, it is important to demonstrate the value proposition of transdisciplinary thinking and transdisciplinary systems engineering in terms of traditional metrics such as cycle time and cost, and desirable system characteristics such as affordability, adaptability, agility, and resilience. As important, is getting social scientists, psychologists, and economists to collaborate effectively with engineers and computer scientists to advance transdisciplinary thinking.

As we look to the future, we have to expand engineering concepts and adapt engineering education for a twenty-first-century engineering workforce. The new breed of engineers will need to acquire a broader skillset. In an era that is being defined by disciplinary convergence, disruptive technologies, and new media delivery platforms, engineers will need a broader set of competencies with the profile being more π-shaped, rather than T-shaped. In other words, the engineer will need depth in a couple of areas (e.g., mechanical engineering and business), with breadth of knowledge in multiple areas. At the same time, academic institutions will have to revise and redefine traditional boundaries inherent in the way today's departments and schools are organized. As important, industry will have to reimagine required competencies based on twenty-first-century systems and ongoing advances in disciplinary convergence. They will need to make sure that engineering education is aligned with these competencies.

These dramatic changes will require both academia and industry to embrace change and adopt new business models. Ongoing convergence among disciplines will continue to open up new avenues for research, even as ongoing technology advances continue to redefine our lifestyle. The growing convergence among disciplines will increase the emphasis on transdisciplinary thinking and the new concepts and opportunities that arise from such thinking. Transdisciplinary thinking will spur the development of new methods, processes, and tools within the rubric of transdisciplinary systems engineering.

Transdisciplinary systems engineering environments will expand to include experiential user interfaces, social media, crowdsourcing, and visual analytics. Model-based methods will converge with interactive storytelling to enhance the exploration of system behaviors in virtual worlds [1]. Model-based interactive storytelling will become the preferred approach for increasing stakeholder participation in collaborative upfront engineering. New theories, inspired by disciplinary convergence, will transform how systems are conceptualized, prototyped, commercialized, and delivered into the marketplace as well as operational environments.

Disciplinary convergence will continue to deepen in the twenty-first century, blurring boundaries between disciplines and causing a dramatic shift in both content (i.e., the material being taught) and instruction (i.e., how material is presented). Entertainment and cinematic arts will become increasingly more interwoven into how science and engineering will be taught and how people will learn in the future. System and data analytics will circumvent the need for making avoidable, possibly erroneous, assumptions. It will become possible to analyze a system's performance potential, test-drive improvements, and justify future capital investments. Experiential interfaces will become a more common place, expanding stakeholder participation in a variety of ways. An appropriately architected system will also be capable of providing embedded training, performance support, and self-directed learning and exploration.

11.2 Keeping an Eye on Technology Advances

Ongoing technological advances will continue to reshape how systems are engineered and how engineering is taught. Some of the more promising advances that will continue to shape engineering and education are presented below.

3D printing, an integral aspect of smart manufacturing, is starting to become an essential element in the classroom for both show-and-tell and hands-on modeling experience. Both engineering instructors and students can be expected to benefit from this direct experience. 3D printers can be expected to play a key role in producing working models to test the efficacy of design principles and heuristics in conjunction with mechanical CAD modeling software. As the cost of 3D printing declines, 3D printers will allow instructors to rapidly and cost-effectively proceed from concept to either virtual prototypes or real-world models in the classroom or laboratory.

Science fiction prototyping (or sci-fi prototyping, for short) is concerned with the exploitation of science fiction to explore alternate futures resulting from the introduction of new concepts and ideas in an evolving system design. The term was coined by Johnson [2], a science fiction enthusiast from Intel, who believes that "science fiction is the playground of the imagination." Johnson advances the proposition that, in theory, it should be possible to design the future with science fiction. Sci-fi prototypes can take the form of short stories, movies, and comics. These prototypes, based on real science and technology, exploit fictional concepts as a key step or an input into the development process. Sci-fi prototypes offer a means to (re) imagine the future in entirely new ways. In light of their versatility, sci-fi prototypes can be expected to become a powerful tool in the hands of systems engineers, designers, system architects, artists, scientists, strategic planners, and, of course, instructors and students. Sci-fi prototypes have already made inroads into AI and robotics. Now they are making inroads into commercial companies and being taught at universities. Inspired in part by Mark Shelly (Frankenstein), Jules Verne (Journey to the Center of the Earth), and H. G. Wells (The Time Machine), the basic idea of Sci-fi prototyping is that *science fiction offers a potential means for scientists to explore the ramifications of new technologies, develop and test hypothesis, and find solutions to problems that invariably arise with the introduction of new technologies and emerging science*. Thus, a sci-fi prototype is, in essence, a short story constructed around a science fact for the purpose of exploring the implications, effects, and ramifications of that science or technology. Sci-fi prototyping provides the context to engage in a meaningful dialogue about the evolution of an existing or envisioned technology. As important, sci-fi prototypes help uncover unintended consequences and facilitate thinking through the ethical implications of an envisioned technology (e.g., self-driving cars). Sci-fi prototyping allows stakeholders to engage in a deeper dialogue about the implications of an envisioned technology. In essence, sci-fi prototypes are a thought experiment and a type of game to assess the legal, ethical, and moral implications of an envisioned technology. Finally, sci-fi prototyping provides a novel lens through which emerging theories can be viewed and explained from different perspectives, even as systems are being developed.

Cloud computing [3] is becoming increasingly popular as its security concerns are being addressed. Cloud computing can essentially virtualize the classroom through online learning platforms. With internet connection, students can log on from anywhere, attend classes virtually, and participate in virtual discussion forums. Homework assignments and tests can be easily disseminated to the virtual class while fostering interaction and discussion. With a 24/7 ubiquitous access to learning resources, students will be able to access instructional material from any location and do homework anywhere at a time of their choosing. In a few years, cloud computing will likely become an integral part of the classroom providing convenient, virtual storage for instructional material (e.g., homework, tests and exams, and lectures).

Augmented reality, a term coined in 1990 by Boeing researcher, Tom Caudell, initially referred to a head-mounted digital display worn by aircraft technicians to guide them in electrical wiring within the aircraft. Today, the term has a broader connotation [4, 5] and is defined as "the interaction of superimposed graphics, audio and other sense enhancements over a real-world environment that's displayed in real-time." It is different from virtual reality in that virtual reality is about replacing the real world with a simulated or virtual world. For example, the superposed "first-down" line seen in television broadcasts of American football games which shows the line that needs to crossed by the team on offense to be awarded a first down is an example of augmented reality. Augmented reality (AR) is a common feature on smartphone apps today. However, while AR is appealing and useful, the experience today is far from seamless. The user still needs to turn on the camera on the smartphone to view results, i.e., it is not sufficiently immersive. Google's Project Glass couples comfortable eyewear with intuitive interactions (e.g., voice commands) and direct presentation of visual information. With a stable connectivity, AR goes beyond what we view. It is, in fact, an effective social media tool for sharing experiences over the internet. Today, AR is making inroads into several sectors including education, navigation (location-based apps), retailing, online shopping, advertising, and video gaming. With today's fast internet connections and vastly improved mobile technology, it is inevitable that social media, email, and news sites will continue to produce a glut of information. So, the next advance driven by necessity, will have to be personalized information presentation. In this regard, next-generation AR will meld information, real world, and AR features in subtle ways to create effective, uncluttered and actionable experiential displays. This fusion will transform human interaction and decision making in the real world as AR dramatically alters the way humans perceive and experience the real world.

Interactive storytelling is a form of digital storytelling in which the storyline is only partially scripted allowing stakeholder users to influence the story evolution through their actions [1, 6, 7]. Interactive storytelling will continue to become increasingly more interwoven in both engineering and engineering education. System stories enhanced through AR, social media, and distributed collaboration will increase stakeholder participation by transforming the experience of upfront systems engineering. For system design, a skeletal storyline will enable stakeholder users to explore system behavior in a variety of what-if contexts, with the results providing support to argue their positions during collaborative decision making. For complex systems, interactive storytelling with augmented reality can be a more effective representation of reality than reality itself. This is because with the information glut produced in a hyper-connected world, stories serve as a contextual filter maximizing the signal-to-noise ratio for each stakeholder in light of their role, concerns, and prevailing context. In education, interactive storytelling contextualizes various concepts, thereby facilitating understanding and recall of key concepts and facts. The basic idea of interactive storytelling is to provide stakeholders what they want in the form they want in the context they want to help with their understanding and decision making.

Telefacturing, a new paradigm in factory operations and smart manufacturing, will become the preferred approach to counter the high costs of factory operations and labor [8]. Telefacturing separates the workforce and the factory. While the factory can be built in locations with manufacturing advantages such as low land prices and tax rate, the workforce can work from preferred metropolitan and urban areas. Workers will be able to remotely control robots and other tele-controllable devices to operate the factory. These capabilities will be realized through state-of-the-art technologies in remote sensing, haptics, remote actuation mechanisms, and telecommunications. Telefacturing will offer several advantages such as dramatic reduction in injuries, human error, and costs. Telefacturing will also reduce human exposure to hazards when devices are remotely operated and processes are remotely executed. Telefacturing will also reduce operating and overhead costs, allowing factories to be located on remote, affordable land with access to electricity and highways, and with modest funds and amenities needed to support onsite engineers and technicians. Ultimately, telefacturing will be environmentally friendly with workers no longer having to make long commutes to remote factory locations. This reduction in commute distance and time will, in turn, reduce the use of fossil fuel and traffic congestion.

11.3 Exploiting Disciplinary Convergence

As technological advances continue and disciplinary convergence broadens and deepens, cross-disciplinary approaches will gradually become part of mainstream engineering and engineering education. Cross-disciplinary approaches, the center piece of transdisciplinary systems engineering, will introduce new perspectives, new insights, and occasionally new concepts into mainstream engineering making complex systems engineering problems tractable. In this book, I have provided examples of how transdisciplinary thinking can help explore, formalize, and exploit the synergy between engineering and other disciplines. This synergy is taking a variety of forms: *enablement* (engineering enables discipline X, or vice versa), *amplification* (engineering enhances or contributes to discipline X, or vice versa), *fusion* (engineering combines with another discipline X to create a new discipline), *emergence* (engineering and discipline X jointly produce a new concept), and *integration* (engineering and discipline X collectively solve a problem that neither could solve alone). As these synergies deepen and produce demonstrable successes, they will add to engineering education content that will be taught in traditional and virtual classrooms. Occasionally, these advances will attract entrepreneurs and investors with the potential of spawning entirely new industries.

Convergent and *divergent* thinking will play complementary roles in disruptive innovation. *Convergent thinking* is concerned with deriving the single right answer to a well-defined question. It emphasizes speed, accuracy, and logic and focuses on recognizing the familiar, reusing/reapplying defined techniques, and accumulating information [9]. It works well for problems with known answers that can be recalled from a library of stored information or deduced from evidence using search, pattern recognition, or decision support strategies. A key limitation of convergent thinking is that it leads to a single best answer with little room for ambiguity. In sharp contrast, *divergent thinking* produces multiple plausible answers from available information. It involves identifying novel combinations of concepts, recognizing associations among distant concepts (nodes), and transforming information into new actionable forms. As important, answers to questions using divergent thinking may vary substantially from one individual to the next. Answers resulting from divergent thinking may be unprecedented and often novel, unusual, or surprising. While both convergent and divergent thinking can lead to new ideas, convergent thinking tends to generates orthodoxy, while divergent thinking generates and fosters new idea creation through variability [4]. The latter is typically associated with creativity. Striking the proper balance between convergent and divergent thinking is the key to realizing creative solutions to complex systems engineering problems.

Transdisciplinary systems engineering in many ways is a natural consequence of disciplinary convergence. It is also being driven by challenges posed by complex sociotechnical systems such as autonomous vehicles, cyber-physical human systems, adaptive energy grids, and smart healthcare. In this book, I have presented several examples of different types of disciplinary convergence that are contributing to the body of knowledge for transdisciplinary systems engineering (Table 11.2).

Table 11.2 Typology of disciplinary convergence (examples)

• Disruptive collaboration: distributed collaboration + social media + crowdsourcing (*fusion*)
• Model-based storytelling: model-based engineering + storytelling + virtual worlds (*fusion*)
• Novel option generation: psychological principles + decision science (*enablement*)
• Human performance enhancement: performance aiding + training psychology (*enablement*)
• Elegant system design: complexity theory + cognitive psychology + systems engineering (*emergence*)
• Affordable resilience: flexible systems modeling + trade-off decision making (*emergence*)
• Autonomous SoS networks: engineering + networking + machine learning + ethics + decision theory (*integration*)
• Twenty-first-century education: storytelling + limited MOOCs + new business models + mobile platforms (*integration*)

There are several other domains (e.g., energy, healthcare) where disruptive convergence is being exploited in creating complex socio-technical systems. In every instance, both the process and the result are contributing to the body of knowledge in transdisciplinary systems engineering. For transdisciplinary systems engineering to become mainstream, the body of knowledge needs to be diffused within academia, industry, and government. A prerequisite for such diffusion are success stories, that can serve as a catalyst and potentially lead to wide-scale adoption and institutionalization of transdisciplinary systems engineering.

11.4 Transforming Engineering Education

The transformation of engineering education will need to be informed by technological advances on the horizon and ongoing disciplinary convergence. The transformation needs to begin with fresh new thinking about engineering education. The transformation also needs to take into account the desires and characteristics of today's learners. And, finally, the transformation needs to exploit new ways of teaching.

New Ways of Thinking About Engineering Education These are exciting times in engineering education. Global competition, emerging technologies and converging disciplines are forcing us to rethink what we teach, how we teach, what combination of technologies to exploit to make learning more effective, and how we apply what we have learned to new problem domains and contexts. Effectively answering these questions will determine how well we are able to produce a twenty-first-century workforce with the requisite competencies in adequate numbers. In other words, the quality of this workforce depends largely on how well we are able to adapt and/or transform engineering education to meet the needs of the twenty-first-century workforce.

Engineering education, which is at the heart of the engineering enterprise, has not kept pace with the needs of the twenty-first-century workforce. This is a concern because attracting and retaining young talent continues to be a national priority. Today engineering education transformation needs to be cognizant of societal, environmental, and sustainability impacts. Unfortunately, engineering education today continues to focus on individual disciplines and narrow technical domains, ignoring the immense potential of cross-disciplinary education enabled by disciplinary convergence. This mindset will clearly need to change if progress is to be made in transforming engineering education. Some of the new ways of thinking about engineering education that needs to inform this transformation follow.

Engineering Education Enterprise is a Complex System. Its transformation needs to be viewed as a complex systems problem that needs to be addressed using methods from complex systems engineering.

Engineering Education Needs to be Broadened. This broadening encompasses the inclusion of certain elements from disciplines such as cognitive psychology, sociology, biology, philosophy, economics, and culture. Broadening engineering education is essential for increasing the value of engineers and their potential employment opportunities. Engineering education needs to embrace innovation and entrepreneurship and nurture creativity. Engineers need to become well-versed in core business principles.

Engineers Must Become System Thinkers. They need to spend time thinking about framing the right problem to solve. They need to think beyond simply generating solutions to solution impacts on the environment, sustainability, and human quality of life. They need to understand the implications of hyper-connectivity and the interdependent nature of the world, in terms of greater likelihood of unintended consequences. They need to understand the need to suboptimize at the subsystem in order to achieve a greater good at the system level. They need to understand that change can propagate in a variety of ways and that some change propagation paths left unchecked can lead to unintended consequences.

Educators Need to Change Perception of Students. Instead of viewing students as passive, empty vessels to be "filled" with knowledge, they should view students as intelligent beings who need to be probed, challenged, encouraged, and supported in their efforts to learn and acquire new knowledge.

Educators Should Instill Importance of Probing Questions and Lateral Thinking. These skills are needed for both problem framing and problem solving. For example, smart questioning offers insights into what a system model needs to represent at a particular point in the system's life cycle to answer questions that arise at that point in the life cycle. Lateral thinking is the key to generating new and novel options and finding appropriate metaphors and analogies to simplify problems and make them more tractable.

Students Need to Understand Social, Geopolitical, and Economic Aspects. Knowledge of these factors adds to their overall value proposition in the job market. It also provides strong justification to graduate engineers in greater numbers with this new skillset. This also means indoctrinating engineers in leadership and social skills, exposing them to a variety of problem situations that may have certain patterns. Today it is imperative that students acquire specific unique skills that are in demand to justify higher salaries relative to other countries.

Storytelling Needs to Become Integral to Instructors in the Classroom. Storytelling offers a means to engage students and foster creative thinking and exploration. It provides the context for information exchange between teachers, learners, and their peers when discussing technical concepts, facts, and other related material. Storytelling needs to be exploited both as a teaching strategy and as a learning approach to enhance reasoning skills and recall.

Calculated Risk Taking and Entrepreneurship Needs to be Rewarded. The ability to transform what has been learned into useful products and services should be recognized as a desirable skill in academia and rewarded accordingly.

New Types of Learners Students today are different in many ways from what they were a decade ago. They are much more technology-savvy, independent thinking, and bolder when it comes to trying out new things. On the flipside, with the ubiquity of smart phones and social media, they tend to have shorter attention spans and are susceptible to distraction. They tend to be more skilled at synthesis, a consequence of divergent thinking. They tend to be diligently engaged around the clock all week with the smart phone being the sole trusted and constant companion and the primary means for engaging with the world around them. They text and tweet at the drop of a hat and have objections to being contacted in this fashion at any time. These behaviors have created a gap between today's students and traditional teaching approaches, as well as between today's learners and instructors.

Teaching approaches and curriculum design need to take these considerations into account. My personal teaching philosophy builds on Winston Churchill's famous refrain, *"I'm always ready to learn, although I do not always like being taught."* To me, reducing this refrain to practice means interspersing lectures with stories (narrative, video) to contextualize concepts, followed by discussions and question answering. This strategy gets students to engage in both critical thinking and systems thinking.

New Ways of Teaching Recent advances in storytelling and massively open online courses are beginning to have an impact on how knowledge is imparted to learners.

Storytelling. Storytelling is an underused teaching strategy, almost a lost communication form. Some engineering educators mistakenly believe that telling stories introduces a needless distraction. Nothing could be farther from the truth. Storytelling can be exploited to draw attention to key points, introduce coherence in seemingly divergent topics, enhance learner recall of facts and principles in context, and effectively break up the monotony of a lengthy lecture.

Storytelling, done effectively, is an effective form of culturally aware teaching [6, 10]. However, effective storytelling is an acquired skill. Storytelling in engineering classrooms can potentially help in shaping the learning experience.

However, most instructors do a poor job of storytelling. They add extraneous details, confuse storytelling with history narration, and fail to engage students in the story.

Fortunately, with advances in technology and social networks, storytelling can be facilitated and become the preferred means to inform, communicate, persuade, and ultimately accomplish learning objectives. However, both instructors and learners need to understand what makes a story effective for teaching and learning. In today's classroom (physical and virtual), when an instructor or a student gets in front of an audience, he/she is competing for attention with social media, and possibly the myriads of apps open on students' laptops. With social media competing for attention, merely lecturing in the classroom is likely to garner scant attention. However, storytelling can overcome that if done right. Doing it right means that stories need to be interactive, attention-grabbing, and memorable in terms of key points and their implications. In learning settings, storytelling needs to be a joint activity with the instructor setting up the story context and initial conditions and the students exploring alternate futures with different assumptions and what-if conditions. This approach allows students to be in control of their learning and engage in critical thinking, a key skill needed in the twenty-first-century engineering workforce. As important, students should be able to introduce "what-if" injects within a story and effectively "rewind" and "playback" the story with different assumptions, technologies, and what-if injects that can take the form of conditions and constraints.

Massively Open Online Courses. Massively open online courses (MOOCs) are free, high quality, and university-level course content delivered over the web. At the present time, learners can enroll in MOOCs to pursue continuing education or undergraduate courses (for credit). The latter can potentially lead to degree programs. With MOOCs, qualified instructors around the world are able to reach learners around the world. It is now also becoming possible to deliver high-quality courses to underserved and remote populations worldwide. MOOCs have the potential to enhance higher education in a financially sustainable fashion.

The first MOOC in 2008 was attributed to two Canadians, George Siemens and Stephen Downes. These researchers were interested in testing their theory of learning via networks. Called "connectivism," this theory advances the proposition that the most effective learning occurs within social networks. Thus, social learning is dynamic and constantly evolving. Unlike knowledge transfer from experts to learners, networked learning is about creation, refinement, and iteration of learning experiences within a community of learners. MOOCs, which create learning experiences by memorializing insights elicited from a community of learners, are called "cMOOCs."

In 2011, MOOCs came into greater prominence when Stanford University's computer scientists, Sebastian Thrun and Peter Norvig, decided to offer MOOCs on artificial intelligence and machine learning, recognizing that all the enabling infrastructure (i.e., video streaming, IT utilities, the internet) was already in place. All they needed was software to stream lectures. They incorporated this software and began streaming their courses over the internet to learners worldwide. The massive open online course (MOOC) became a reality. Approximately 160,000 people worldwide enrolled in their AI course. The extraordinary reach of the internet was being fully exploited. Some began to view MOOCs as a technology advance that would render traditional classrooms obsolete. With learners being able to log on from any location to access streamed lectures, the concept was liberating and irresistible.

Expectedly, other universities began to videotape lectures and make them available over the internet. MIT offered OpenCourseWare, while Yale offered Open Yale. This broadcast form MOOC, called "xMOOC," signified massive scalability well beyond what online courses and in-person learning could provide. The early market entrants (i.e., Coursera, Udacity, EdX) into this space garnered substantial financial backing. They were viewed as change agents who were exploiting technology to reinvent education.

However, this story did not quite turn out the way it was being imagined [11]. Learner completion rates became an issue. Initially, the average completion rate was less than 7%, subsequently rising to 12.6%, with considerable variation from course to course. MOOCs had clearly underperformed expectations. The question was "Why?" Upon closer examination, it became apparent that it takes more than technology to achieve desired educational outcomes. The initial MOOC largely ignored learner proclivities and educational trends. Some learners, without "skin in the game" (i.e., free courses), turned out to be "grazers" who upon locating what they were looking for, discontinued with the course. Others, who enrolled because the course was free, had no real commitment to completing the course and left after watching a few minutes of the lecture. Those who completed the course turned out to be individuals with advanced degrees.

It soon became apparent to MOOC developers that the knowledge of computers, networks, and subject domain was not ever going to be enough, if learner expectations and proclivities and effective learning strategies were not reflected in the learning experience. That meant that it took more than computer science to create the right learning experience. More specifically, it meant that reliable video recording of lectures, multiple choice quizzes, and unstructured discussion forums that the initial generation of MOOCs offered were necessary but not sufficient to create the right learning experience. What was missing in the initial MOOCs was an understanding of the learner's behavior and expectations and the use of effective pedagogical interventions. In other words, it was going to take a combination of computer science, learning science, and cognitive science to develop an effective MOOC.

- *Computer science*: for video streaming, networks, IT infrastructure
- *Learning science*: for active learning strategies (i.e., learning through peer-to-peer interaction, virtual teamwork, interactive exercises)
- *Cognitive science*: for intuitive human-computer interaction, while being mindful of human cognitive limitations

The impact of active learning for science, engineering, and mathematics is well-documented in a 2014 report published in the *Proceedings of the National Academy of Sciences*. Two significant findings from this report were that average test scores

Table 11.3 Benefits and challenges of MOOCs

- Benefits
  - modest tuition
  - high-quality content and education regardless of location
  - can coexist with and complement bricks-and-mortar establishment
- Challenges
  - student authentication is a problem with all online courses – need proctored testing centers, biometrics
  - poor completion rates (7–8%)
  - difficult to gauge credit hours earned
  - automated assessment (multiple choice, matching, fill-in-the-blanks) with teaching assistant-in-the-loop (limited class size)
  - draw student away from traditional brick-and-mortar establishments
  - require creation of new, sustainable business model that complements bricks-and-mortar establishments
  - roles of students, teaching assistants, and faculty members may have to change

went up approximately 6% with active learning, and students taking traditional lecture classes were 1.5 times more likely to fail than those who engaged in active learning.

More recent MOOCs have introduced online learning innovations such as active learning. MOOCs will continue to give millions across the globe access to a wide range of topics but now with effective pedagogical interventions and innovations incorporated in course delivery. Despite these advances, it is highly unlikely that MOOCs will replace successful online degree program. However, they will become entirely viable as yet another digital learning option. And, for those who not interested in pursuing a degree program, MOOCs offer an attractive option to acquire new knowledge.

In light of the foregoing, the overall impact of MOOCs cannot be denied. Today millions across the globe have access to courses presented by scholars from reputable universities. Even the low 7% completion rate still translates to thousands of students completing courses based on enrollment data on MOOCs. In fact, this is significantly more than the cumulative student enrollment in most colleges [11] .

MOOCs currently employ one of two models: (a) web-based and emailed course content, with assessment through automated exams (e.g., MOOCs offered through EdX), and (b) "connectionist" learning that has less structure and content and in which learning occurs through social interaction methods such as blogs, threaded discussion boards, and email. In both cases, teaching assistants typically moderate interactions and answer questions. MOOCs differ from traditional distance education networks in two important ways. One way is through *worldwide massive enrollments* with high-quality curricula available *to all* in *any location* with internet access. Another is that they offer a *cost-effective alternative* for regions where *brick-and-mortar* facilities are *infeasible*. Table 11.3 presents the benefits and challenges posed by MOOCs.

Despite their obvious benefits, there is palpable reluctance on the part of students, instructors, and universities to fully embrace MOOCs. It is expected that most universities will participate in MOOCs to a limited extent (i.e., for a select few credit and non-credit courses) or grant limited waivers (e.g., waive prerequisites) to those who complete MOOCs. However, current data suggests that this is not likely to be a widespread phenomenon.

But what if MOOCs are shown to be effective for mainstream learning. If this happens, then they have the potential to be truly transformative. For example, a well-known quality metric for traditional universities is the average number of students per class, with a lower ratio viewed as more desirable. With MOOCs, this metric will cease to be relevant, because automated course delivery and grading allows for dramatic increase in course enrollments. Does the proliferation of MOOCs mean that fewer faculty members and more teaching assistants will be needed in the future? Only time will tell. In the meantime, there is mounting pressure on traditional universities to increase class sizes by adopting some MOOC features (e.g., automated grading), while preserving instructor-student interaction.

There are potential unintended, undesirable consequences with MOOCs especially if students' performance on courses, and their grades, current jobs, and interests, get in the hands of placement organizations, real estate agents, and the like. Students could then become easy, unwitting targets of aggressive marketers.

There are lingering concerns with MOOCs in their current form. First, MOOCs do not offer an ideal learning environment. They continue to offer minimal peer-to-peer interaction and minimal interaction with mentors. This is entirely understandable because these capabilities are real cost drivers! Second, learning is a social phenomenon that needs to occur in social settings, not in isolation. Third, learning is inseparable from the "experience of learning" and knowledge creation. Therefore, MOOCs have to provide a means to satisfy a fundamental human need – the social experience. Finally, learning needs to be learner-centric and pedagogically based if they are to succeed in the marketplace.

Optimizing Teaching Resources and Quantifying Return on Learning Organizations across the globe continue to make substantial investments in training and development [12]. At the present time, these corporations have no way to optimize investments in the available learning options and assess their impact on employee learning proficiency metrics (e.g., time-to-proficiency). With fixed budgets, they have a pressing need to allocate both funds and cohorts optimally to different learning options (e.g., online learning, in-person instruction) available at different geographic locations subject to location constraints. Another shortcoming today is the inability to address specific learning deficiencies with targeted training. As a result, companies currently resort to "brute force" mass training, thereby substantially driving up training costs. Finally, organizations have no way to quantify the return on their investment in workforce training, learning, and development.

To address these shortcomings, a *learning curves*-driven approach was developed [13]. The central idea behind this approach is that accumulating experience leads to enhanced performance. Learning curves describe learning rates both in production environment as well as workforce training environments [14]. The *learning curve* concept was originally discovered by the Boeing Company and introduced in aircraft production in the mid-1930s [15–19]. Boeing discovered that the time taken to assemble an individual aircraft declined with an increase in the number of aircraft assembled [20–22]. This finding revealed that the cost to assemble a new airplane was highly predictable. This finding was also supported in workforce training, where the rate of improvement or increase in learning speed exhibited sufficient repeatability for characterization by a mathematical formula. The learning curve phenomenon for workforce training was tested and found to hold across different industries, with the only variation stemming from differences in learning rates across industries.

Organizations faced with the decision to invest in different training options are typically concerned with two key metrics: time spent on training (*t*) to reach proficiency (*P*) and return-on-learning (*RoL*) measured in terms of proficiency development rate (individual/workforce) and profitability. The learning curve, and more specifically, the proficiency curve embodies three key principles: (a) it is possible to map learner characteristics to appropriate learning options; (b) time-to-proficiency decreases with proper matching of learners to learning options and optimal allocation of the budget (a fixed amount) to the different learning options; and (c) reduction in time-to-proficiency follows a predictable, repeatable pattern. A proficiency curve depicts the time-to-proficiency for a given amount of time (or money) spent on specific learning options. For a desired level of proficiency, it is possible to identify when that proficiency level is satisfied.

Historically, a learning (proficiency) curve was used to *describe* the state of learning of an individual or cohort. However, Madni et al. [13] employed the learning curve as a function to *optimize* given resource constraints and learner "forgetting" behavior. Forgetting occurs when, for example, classes are not offered over a weekend, or learning is disrupted by interruptions such as temporary reassignment of learners to jobs that prevent their ability to continue learning on the job. The ability to optimize a learning curve with resource/budget constraints and a defined set of learning options has wide applicability for major corporations, ranging from production engineering, manufacturing operations, supply chain management, and customer-facing functions in industry.

Learning curve optimization (LCO) is inherently a resource allocation problem in which (a) a fixed resource (e.g., budget or time) needs to be optimally allocated to available learning options and (b) individual learners (or cohorts) are optimally assigned to different learning options. The end goal of LCO is to accelerate learner/cohort time-to-proficiency. It is important to realize that while knowing the amount of time or money spent on workforce learning is valuable, it is but part of the story. Understanding the return on investment (ROI) in learning completes the story.

Budget and learner allocation problems are a generalized form of the *portfolio optimization problem*, in which the total investment across *n* investment (i.e., training) options is optimized by assigning most appropriate weights (i.e., fraction of budgeted dollars) for each option. The optimization problem is concerned with maximizing a particular portfolio measure defined by the investor, typically *portfolio return*.

Table 11.4 Techniques used in learning curve optimization

• **Curve Fitting**: statistically fit curves to data collected for various learning options (e.g., in person, online)
• **Portfolio Optimization**: optimize resource allocation to maximize learning effectiveness; traditional constrained optimization problem is adapted to incorporate employee workforce characteristics (e.g., language proficiency, technical knowhow)
• **System Dyamics**: (a) the world consists of dynamic, interconnected, sociotechnical systems; and (b) there are tools that can be used to further our understanding of these dynamic relationships
• **Agent-based Modeling**: characterize progress of learners and groups of learners (cohorts) as they take online or in-person courses [10]; well-suited to capturing known interactions between instructor and learner, between learners, learner and content, and learner and training modality; also useful for uncovering previously unknown interactions among agents

The training investment allocation problem deviates from the traditional portfolio optimization problem in a couple of key dimensions. First, it does not include the value of the training for each training option. For that, we need to consider some performance measure associated with employees trained. Nor does it include the unique personality and learning traits of the individuals trained. The current training approach is a "one size fits all" approach. Furthermore, even though enormous emphasis tends to be placed on diversity in the workplace in most organizations, this is not the case when addressing in-house education. It is important to recognize that the concept of diversity goes beyond language and culture to the different ways in which people learn. Education alone will not necessarily produce the knowledge needed to perform a practical task successfully. Hands-on experience is often necessary to exhibit certain skills and competencies. Therefore, we assume some prior (fixed) knowledge about the effect of each training area on the traditional knowledge, skills, and abilities (KSA) metrics. Then, associated with each learning option are measures of both intellectual and practical proficiency gained from each learning option.

In light of the foregoing, the solution to the learning curves optimization problem draws on and combines techniques from multiple disciplines [14]. Specifically, the approach exploits curve fitting (statistics), portfolio optimization (business), system dynamics (systems theory), and agent-based modeling (computer science). Each technique is presented in Table 11.4.

11.5 Challenges Ahead

The twenty-first century will continue to see advances in disciplinary convergence on two fronts: microsystems (nano, bio, info) and macrosystems (energy, environment, healthcare, manufacturing, communication, and logistics). Globalization will continue in R&D, innovation, and manufacturing sectors [23]. The open innovation trend can be expected to continue in the foreseeable future. Just consider that Boeing

787 reportedly has 132,500 engineered parts which are produced in 545 locations. Open US campuses with traditional immigration policies have served us well in the past with most international engineering Ph.D. students, primarily from Asian countries, deciding to stay in the USA and contributing to the US workforce [23]. That has changed. Now these students are returning to their home countries in increasing numbers because attractive jobs and start-up opportunities await them in their home countries. In light of these trends, Vest [23] recommends that we need to make attending US universities and engineering schools "an exciting, creative, adventurous, rigorous, demanding, and empowering experience." This is especially true today because today's learners are quite different from what they were a decade ago. They are much more technology-savvy, independent thinkers, and quite bold when it comes to trying out new things. Accustomed to the use of social media and smart phones, they tend to have shorter attention spans and are early distracted. They tend to be skilled at synthesis, a consequence of divergent thinking. The smart phone has become their constant and trusted companion and the primary means for engaging with the outside world. They text and tweet at the drop of a hat and are not averse to being contacted in this fashion at odd hours. These behaviors have led to a noticeable gap between today's learners and today's teaching approaches, as well as between today's learners and instructors. The problem is further complicated by the fact that engineering educational transformation will have to contend with financial constraints along the way. Fortunately, new methods for quantifying return on learning have emerged that are intended to help make each dollar go farther [13].

Ultimately, convergence and transdisciplinarity are more than just about engineering. Norm Augustine [24] poses the question whether a life can be "truly fulfilling with only knowledge of magnetohydrodynamics, quantum mechanics, and matrix algebra and no exposure to the beauty and utility of Beethoven, van Gogh, and Shakespeare." He emphasizes that there is much more to be gained from such works than just aesthetic beauty. In a coauthored book, *Shakespeare in Charge* [25], he and Adelman put forth the proposition that by understanding Shakespeare, one can better understand business. They back up this claim with a simple explanation: "Business is about people, and no one understood people better than Shakespeare did." I couldn't agree more.

Thus, transdisciplinarity and convergence is not just about enriching engineering and making the intractable tractable. It is about living a more fulfilling life, with greater awareness of what becomes possible when disciplines collide, harmonize, and give rise to new insights.

In light of these trends, we need to pursue opportunities on two fronts: *cross-disciplinary research* that exploits disciplinary convergence and contributes to the maturation of transdisciplinary systems engineering and *engineering education transformation* to reflect these new advances in both educational content and teaching approach. This transformation will make attending US universities and seeking jobs in the US attractive for international students again.

We are just beginning this journey. For those inclined to conduct research and contribute to education in transdisciplinary systems engineering, there could not be a better time.

References

1. Madni AM, Nance M, Richey M, Hubbard W, Hanneman L. Toward an experiential design language: augmenting model-based systems engineering with technical storytelling in virtual worlds. Procedia Comput Sci. 2014;28:848–56.
2. Johnson BD. Science fiction prototyping: designing the future with science fiction. Synth Lect Comput Sci. 2011;3(1):1–190.
3. Hassan Q. Demystifying cloud computing. J Def Softw Eng. 2011; 16–21.
4. Azuma RT. A survey of augmented reality. Presence Teleop Virt. 1997;6(4):355–85.
5. Azuma R, Baillot Y, Behringer R, Feiner S, Julier S, MacIntyre B. Recent advances in augmented reality. IEEE Comput Graph Appl. 2001;21(6):34–47.
6. Madni AM. Expanding stakeholder participation in upfront system engineering through storytelling in virtual worlds. Syst Eng. 2014;18(1):16–27.
7. Madni AM, Richey M, Ordoukhanian E, Venkatesh J, Zender F, Chang K, et al. Exploiting storytelling in collaborative systems engineering: Towards a smart experiential dashboard. Conference on Systems Engineering Research Hunstville, Alabama, 2016.
8. Khoshnevis B. Telefacturing – a paradigm for the 21st century. Ind Eng. 2015;47(11).
9. Cropley A. In praise of convergent thinking. Creat Res J. 2006;18(3):391–404.
10. Madni A. Integrating humans with and within software and systems: challenges and opportunities. (Invited Paper) CrossTalk, J Def Softw Eng. 2011;5:4–8.
11. Ubell R. How the pioneers of the MOOC got it wrong. IEEE Spectrum. 2017.
12. Richey M, Nance M, Hanneman L, Hubbard W, Madni AM, Spraragen M. A complex sociotechnical systems approach to provisioning educational policies for future workforce. Procedia Comput Sci. 2014;28:857–64.
13. Madni AM, Paulson C, Spraragen M, Richey MC, Nance ML, Vander Wel M, editors. Model-based optimization of learning curves: implications for business and government. INCOSE international symposium. Wiley Online Library; 2015.
14. Madni AM, Sievers M. Model based systems engineering: motivation, current status and needed advances, Technical Report. University of Southern California, Viterbi School of Engineering, Systems Architecting and Engineering. 2015.
15. Adler PS, Clark KB. Behind the learning curve: a sketch of the learning process. Manag Sci. 1991;37(3):267–81.
16. Anderson EG, Parker GG. The effect of learning on the make/buy decision. Prod Oper Manag. 2002;11(3):313–39.
17. Argote L. Organizational learning curves: persistence, transfer and turnover. Int J Technol Manag. 1996;11(7-8):759–69.
18. Argote L. Organizational learning: creating, retaining and transferring knowledge. Berlin: Springer Science & Business Media; 2012.
19. Epple D, Argote L, Devadas R. Organizational learning curves: a method for investigating intra-plant transfer of knowledge acquired through learning by doing. Organ Sci. 1991;2(1):58–70.
20. Wright TP. Factors affecting the cost of airplanes. J Aeronaut Sci. 1936;3(4):122–8.
21. Yelle LE. The learning curve: historical review and comprehensive survey. Decis Sci. 1979;10(2):302–28.
22. Zangwill WI, Kantor PB. Toward a theory of continuous improvement and the learning curve. Manag Sci. 1998;44(7):910–20.
23. Vest CM. Educating engineers for 2020 and beyond. National Academy of Engineering. 2005.
24. Augustine NR. One cannot live by equations alone: education for life and work in the twenty-first century. Lib Educ. 2013;99(2):n2.
25. Augustine NR, Adelman KL. Shakespeare in charge: the bard's guide to leading and succeeding on the business stage. Hyperion-Talk-Miramax; 1999.

Appendix A: Glossary of Key Terms and Definitions

Accident	An undesirable, unintended event that produces an economic loss and/or a loss of function and/or a loss of human lives
Adaptability	Ability of a system to internally change to continue to fit within a changing environment; includes self-modification and self-organization
Agility	Ability of a system to rapidly and cost-effectively undergo change; flexibility is a prerequisite to agility
Ambiguity	Potentially associated with several possible meanings or interpretations
Analogy	A comparison between two entities on the basis of their structure for purpose of understanding, learning, explanation, or clarification
Architecture	A formal description and representation of a system, organized in a way that enables reasoning about the structure, behavior, and interfaces in a system
Assumption	Something that is accepted as true or certain to happen, without proof or evidence
Augmented Reality	A live (direct or indirect) view of a physical, real-world environment whose elements are augmented by computer-generated sensory input such as sound, video, graphics, or GPS data
Autonomous Vehicle	A vehicle that is capable of sensing its environment and navigating without human input for extended durations
Belief State	Belief in what the state of a partially observable system is; a term associated with POMDP and other probabilistic modeling approaches
Cloud Computing	The use of a network of remote servers hosted on the Internet to store, manage, and process data, rather than a local server, or a personal computer

A.M. Madni, *Transdisciplinary Systems Engineering*,
DOI 10.1007/978-3-319-62184-5

Complex System	A system in which the interactions, interconnections, or interdependencies among components make it difficult to understand, model, design, manage, and change the system and predict its behavior
Concept of Operations	A user-oriented document (occasionally supplemented with a prototype) that describes the characteristics of an envisioned system from a user's perspective. A CONOPS also describes the mission, operational setting, and objectives from the perspective of the different stakeholders in quantitative and qualitative terms
Contract	A formal, "assert-guarantee" approach for requirements engineering that is also used for virtual integration and deployment; used in both software and systems engineering
Contract-Based Design	A process-neutral approach that qualifies each design step with regard to system integration. Therefore, it is a compliant with various SE process models, including the V model
Crowdsourcing	The activity of obtaining needed services, ideas, or content by soliciting contributions from a large group of people, especially an online community, rather than from traditional services (i.e., employers, suppliers)
Cyber-Physical Systems	Engineered systems that are built from, and depend upon, the seamless deep integration of computational algorithms and physical components
Decomposition	A divide-and-conquer approach that breaks down a system into its components to facilitate analysis conducted at the functional, logical, and physical levels of the decomposition hierarchy
Decision Theory	A body of knowledge and related deterministic and probabilistic techniques of different degrees of formality designed to help decision makers choose from an available set of alternatives in light of their possible consequences. Decision theory applies to conditions of certainty, uncertainty, and risk. Decision theory is beginning to draw on other disciplines such as psychology, philosophy, and systems thinking
Disciplinary Convergence	The merging of humanities' ability to expand on and theorize with computer and information science's ability to design efficiently and make meaning implicit in the design and code
Disruptive Collaboration	Working together remotely as a team, sharing data and partial solutions through online and mobile networks to create shared value. The collaborators tend to be known to each other

Distributed Collaboration	The convergence of distributed collaboration, social media, and crowdsourcing to expand the universe of contributors to problems in need of solutions; changes the definition and collaborators
Effectiveness	Ratio of performance of functions achieved and totality of functions and performance desired
Efficiency	Ratio of performance of functions achieved and resources used
Elegant Design	A design that does away with having to make performance and capability limiting trade-offs among desired system characteristics; a design that offers the requisite functionality with minimum structural complexity
Emergence	A system-level property that is the result of latent system properties and behaviors that are discovered, those that show up spontaneously over time and space, and those that arise as a result of interactions with other systems and the environment
Enterprise	A value-generating, profit-making organization that returns value to all stakeholders as a result of interactions and contributions of its members
Evolvability	Ability of a system to readily change with the emergence of new needs, requirements, and technologies over time
Experiential (User) Interface	A user interface in which the experiences are specifically created by the UI designer to fully engage users using a variety of multisensory cues and metaphors
Extensibility	Degree to which a system or its components can be extended to a higher level of abstraction or to introduce a new capability
Flexibility	Ability of a system to be changed in specific dimensions in a variety of ways
Formal Methods	System design techniques that use rigorously specified mathematical models to verify a complex system's properties in a more thorough manner than empirical testing
Functions	A specified or desired system behavior that is repeatable and that is associated directly or indirectly with the stated purpose of the system
Heuristics	A mental shortcut or "rule of thumb" that enables rapid solution of complex problems most of the time
Innovation	The translation of a novel idea or invention into a good or service that creates value for which customers are willing to pay
Integration	Actively associated with making required system components work together, to satisfy system requirements

Interconnection	Relationship between physical or abstract entities with the connection serving as pathway for interaction
Interaction	A property of entities that exchanges information, energy, or matter that crates or contributes to their independence
Interdependence	Relationship between any two entities that cannot exist or operate without the other
Interdisciplinary	Combining or involving two or more disciplines or fields of study to respond to needs that cannot be addressed solely by a single discipline
Interface	The point where two systems, people, and organizations interact with one another
Internet of Things	Smart connected devices and structure with embedded algorithms, software, sensors, actuators, and with network connectivity that allows these objects to collect and exchange data; an enabler of cyber-physical systems
Interoperability	Ability of different IT systems and software applications to communicate, exchange data, and exploit the exchanged information
Interactive Storytelling	A form of digital entertainment in which the storyline is not predetermined and can be shaped by users as they interact with elements in the story
Lateral Thinking	A mental activity to generating options and solving problems that employs nontraditional, indirect horizontal thinking instead of the "vertical" or logical thinking employed in traditional problem solving
Learning Curves	A graphical depiction of the increasing in learning (vertical axis) with experience (horizontal axis)
Learning Organization	An organization that acquires knowledge at all levels to accelerate innovation to survive and thrive in a rapidly changing global business landscape
Metaphor	A figure of speech which makes an implicit, implied, or hidden comparison between two unrelated concepts that share some common characteristics
Modularity	Degree to which system components can be designed, built, operated, and swapped without impacting other components
Module	A part of a system that is built to have minimal and standardized interaction with the rest of the system
Model-Based Systems Engineering	A computer-supported engineering approach that represents a shift from document-centric to model-centric system development
Model-Based Storytelling	The mapping of system models to virtual worlds in which system stories are interactively executed by stakeholders

Operations Research	A theoretically grounded, formal approach to executing decision-making that harnesses methods from system modeling, problem formulation, mathematical modeling, and system modeling
Optimization	An analytic approach for maximizing the performance of a system subject to constraints
Option Generation	A central activity in decision-making concerned with the production of alternatives (known, newly created) using a variety of psychological principles such as constraint relaxation and lateral thinking
Partially Observable Markov Decision Process	A generalization of a Markov decision process that models an agent's decision process in which system dynamics are determined by the MDP, but the agent cannot directly observe the underlying state. Therefore, the agent maintains a probability distribution over the set of possible states, based on the set of observables and observation probabilities, variety of real-world problems. Well suited for robot navigation process and planning under uncertainty
Platform	A business strategy that relies on network effects and "pull." It leverages three transformative technologies: cloud, social, and mobile. The cloud provides a global infrastructure for production of social networks which connect people globally and maintain their identity online. Mobile allows connection to this global infrastructure anytime, anywhere. The result is a globally accessible network of entrepreneurs, producers, and consumers who are available to create business, create content, and sell and purchase goods and services
Quality Attributes	System properties associated with nonfunctional requirements; examples are maintainability, testability, and flexibility
Query Reformulation	A strategy for refraining how a question is asked to illuminate new perspectives about a problem/system. This strategy is effective for requirements definition in system development
Repurposing	The activity associated with adapting a product for a new purpose or to satisfy a different need
Resilience	Ability of a system to absorb and adapt to disturbance, description, or change that falls outside the base adaptive mechanisms/model defined in the system. This definition distinguishes resilience from adaptability, which is the ability to absorb or adjust to perturbations (i.e., disturbance, disruption, or change)
Return on Learning	The benefit resulting from investments in a portfolio of learning/training strategies. The benefits are in the form of time to competency and rate of workforce increase

Reuse	The activity associated with repeated use or application of the same module or component in different functions of a system, in different systems, or at different times in the same system
Risk	The product of the likelihood of an accident or loss event and the severity of the accident or less event
Robustness	Ability to sustain performance under a variety of perturbations (internal or external)
Scalability	Ability of a system to maintain performance, function, and quality attributes with an increase in scale without a corresponding increase in system's complexity; does not include an increase in scope usually
Science Fiction Prototyping	The use of science fiction to describe and explore the implications of futuristic technologies and the social structures enabled by them
SCO	A launchable learning object that communicates with the run-time environment that launched it; can be purposed for both learning and performance support
SCORM	A collection of standards and specification for e-learning; defines communications between client side content and the learning management system ("run-time environment")
Social Network	A social structure made up of a set of actors (i.e., individuals or organizations), sets of dyadic ties, and social interactions between actors. The social network perspective provides a set of methods for analyzing the structure of whole social entities as well as a variety of theories explaining the patterns observed in these structures. The study of these structures uses social network analysis to identify local and global patterns, locate influential entities, and examine network dynamics
Sociotechnical Systems	A complex system comprising human and machine elements in which social and management dimensions are significant, and the interactions among the elements produce value for the stakeholders
System	A set of interacting components with a defined purpose or behavior
System of Systems	A predefined or dynamic assembly of dedicated systems that pool their resources and capabilities to create a new system with capabilities beyond the sum of capabilities of the individual systems
Systems Architecting	An integrative, trade-offs-driven process for creating high-level conceptual structure, interfaces, and behaviors of a system that conforms to standards and interface protocols in satisfying customer and stakeholder requirements

System Architecture	A conceptual model that defines the structure, behavior, and other perspectives of a system
Systems Engineering	An interdisciplinary approach and means to enable the realization of successful *systems*. Successful *systems* must satisfy the needs of their customers, users, and other stakeholders
System Dynamics	An approach to modeling complex systems in the social sciences and management that employs network of nodes and feedback relationships among them
System Modeling	Vocabularies, symbols, rules, and representations (behavior, structure) that make use of vocabularies, symbols, and each of the purposes of displaying and predicting systems structure and behavior
Systems Thinking	Ability to think holistically and in terms of trade-offs, judgments, with the recognition that latent system properties and interactions could lead to emergent behavior
Swarm	A collection of simple agents interacting locally with each other and the environment in accord with simple rules and without a centralized control structure dictating how each agent should behave. The local and, to a certain degree, random interactions among agents lead to emergent global behavior, unknown to the participating agents
Technological Convergence	The evolution of technology systems to evolve toward performing similar tasks by merging existing technologies into new forms that harness different media and applications
Traceability	The ability to verify the history and source of a requirement in system development
Trade-off	A balance achieved between two or more desirable, competing properties through reasoned compromise among stakeholders
Trade-off Analysis	An activity concerned with answering a basic question: are solutions being suggested as good as possible – are they on the Pareto frontier? How much must I give up to get a little more of what I want most?
Uncertainty	A state of knowledge that arises from having imperfect incomplete information about a system and/or its environment as well as unknown system states arising from partial observability of the system
Utility	A measure of preferences over a set of goods/services; an important concept in rational choice theory in economics and game theory
Validation	An activity performed at the end of the testing cycle to determine how well a product or service addresses the business needs that hopefully are reflected in the requirements specification – "did we build what the market needs?"

Value	A measure of the benefit provided by a good or service to an economic agent; generally measured relative to units of currency, and the interpretation is therefore "what is the maximum amount of money a specific actor is willing and able to pay for the good or service"? Value can be either objective or subjective. In the latter case, it means subjective worth or importance. For example, "the value of future benefits to the decision maker," "the value of clean air to the society." For the purposes of analysis, the subjective values must be measured on some scale. These measures of value should be based on preferences expressed by the person or group of interest
Verification	Testing to determine whether or not a product or service satisfies its specification – "did we build what we signed up to?"

The manufacturer's authorised representative in the EU is Springer Nature Customer Service Centre GmbH, Europaplatz 3, 69115 Heidelberg, Germany. If you have any concerns regarding our products, please contact ProductSafety@springernature.com

Printed and bound by CPI Group (UK) Ltd, Croydon, CR0 4YY
30/09/2026
02225593-0001